When we have discussions on sustainability, we need to remind ourselves that we are called to be good stewards of the resources the Lord has blessed us with. Psalms 23:1 says: "The Lord is my Shepherd, he hath provided everything I need." I challenge each one of us to be those chosen stewards. Rick's book is biblically based and written from a Christian perspective.

—**Bob Dixson**
Mayor of Greensburg, Kansas

The writer of Proverbs wrote, "There is a way that seems right to a man, but in the end it leads to death." Centuries later, Reinhold Niebuhr wrote, "The trustful acceptance of false solutions for our perplexing problems adds a touch of pathos to the tragedy of our age." Man too often only listens to himself. Now Rick Gasaway suggests we might benefit by listening rather to the Creator of life and the world we occupy. We might then get direction and insight into our selfishness and see that by putting others first we take care of the problems humans encounter on their own. Here is an educating, although sometimes deep for the less technical, challenging call to a change in thinking as we seek to be good stewards of the world we call home for only a short time. A must read for those who are serious about leaving this world a better place because they lived here.

—**Pastor Ken Sype**
Overland Park, Kansas

Rick Gasaway articulates in clear, compelling language the ways in which we all need to band together to solve our most pressing environmental problems. While I do not agree with every premise, I laud his courageous effort to thoughtfully bridge the environmental divide and acknowledge the powerful role of faith in forging a cleaner, greener energy future.

—**Simran Sethi**
Lacy C. Haynes Professional Chair
University of Kansas School of Journalism and Mass Communication, Lawrence, KS

An
Inconvenient
Purpose

Linking Godly Stewardship and Alternative Energy

Alex,

I'm looking forward to
working with you and
the GenCell team on fuel cells.
Enjoy the book!

Richard Gasaway

God bless,
Rick Gasaway

WinePress **WP** Publishing

Gen 2:15

WinePress Publishing (PO Box 428, Enumclaw, WA 98022) functions only as book publisher. As such, the ultimate design, content, editorial accuracy, and views expressed or implied in this work are those of the author.

Unless otherwise noted, all Scriptures are taken from the *Holy Bible, New International Version*, NIV. Copyright © 1973, 1978, 1984 by the International Bible Society. Used by permission of Zondervan. All rights reserved.

ISBN 13: 978-1-60615-009-2
ISBN 10: 1-60615-009-X
Library of Congress Catalog Card Number: 2008943016

To my Lord and Savior, Jesus Christ,
and my patient wife, Bonnie,
my amazing son, Garrett,
my great friend and brother in Christ, David Pauli,
my wonderful parents,
Alexis Kling,
Kate Wiens,
Pastor Ken Sype,
Heart of America Christian Writer's Network,
Ron Wheeler,
and all the others who helped me on this journey.

CONTENTS

ILLUSTRATIONS

Figures

PREFACE

E ven a blind squirrel finds a nut every now and then. I often use that expression in jest (just ask my wife), but it describes how I feel about the tidbits of wisdom coming out of both liberal and conservative camps regarding the true human impacts on global warming and our insatiable demand for energy. You see, in my profession as an electrical engineer, I straddle a world between the traditional, incumbent views of energy supply technologies and the radical need to overhaul our energy supply paradigm, thus changing the world as we know it. I do know one thing, however: whatever we do as an energy-consuming global society, the economy will follow in lockstep, and the environment will follow in tow. Here's the question that demands an answer: *What is the wise path to sustainable alternative energy solutions and a healthy environment?*

Like many, I experience the polarization of both conservative and liberal camps firsthand. Attending one meeting, I see frustration build in some global warming believers. They cannot fathom why some people deny the irrefutable truth of human-caused global warming. In their minds, scientific consensus has been reached; the case is closed. I pray nobody utters that four-letter word, "Bush," as in George W. Bush, or the whole meeting may sink into the quicksand of pent-up political frustration.

The next week, in a meeting discussing energy policy with a group of conservative Republicans, a crescendo of frustration builds on mandated environmental protectionism. Inevitably, somebody invokes a different four-letter word, "Gore," as in Al Gore, and the feeding frenzy of anti-liberalism begins.

In each meeting, the participants expressed complete conviction that their side is noble and irrefutably correct. Each side's cancerous attitude towards "them" disturbed me since I have good friends on both sides of the fence. The dismissal of "those environmental destroyers" or "those radical environmentalists" does little to foster an attitude of constructive stewardship. Conservatives and liberal progressives each consider the other foolish, unrealistic, and just flat wrong, to put it gently.

The Christian apologist writer G. K. Chesterton, put it another way, "The whole modern world has divided itself into Conservatives and Progressives. The business of Progressives is to go on making mistakes. The business of the Conservatives is to prevent the mistakes from being corrected."[1] How do we move past that?

The answers reside within Holy Scripture. The Bible contains all the answers about how to treat each other and our environment. Many think the science involved for energy and environmental problems would preclude our use of the Bible for guidance, but they are wrong. Granted, you won't find global warming, fuel cells, or many other scientific phrases mentioned specifically in the Bible. But Scripture teaches us how to live as Christians. It therefore gives us answers on how to *apply* technology and science to our lives. The Bible is still the best "how to" book around.

Looking ahead, the key steps to constructing a world powered by alternative energy rely on a biblical worldview and the discovery that energy usage trumps climate change, economic development, and politics in importance. They are all related, but energy usage is the root cause of the problems; the rest are symptoms. In medical terms, curing the root cause is better than just treating the symptoms. Our chance to solve energy problems lies before us, while the rest of the world continues to treat the symptoms. I doubt many Christians have biblical stewardship of energy issues high on their priority lists, but that is changing. It is an inconvenient purpose placed in our hearts as Christians.

If energy is a problem, what are the answers? We cannot stop using energy, but we cannot continue the unsustainable pace at which we live now. For example, we cannot stop all the coal trains and shutter the existing coal-fired power plants. Coal provides too much of our electricity (about 50% in the U.S.) to change it drastically overnight. However, that doesn't mean we shouldn't try to significantly scale back, or modify coal-fired electricity.

Wind, solar, fuel cells, tidal, bio-fuels, and other alternative energy technologies hold promise for the future and should enjoy great expansion. However, these alternative technologies can only expand so much until we resolve intermittency issues, reliability issues, power quality issues, environmental impacts, and economic restraints.

Nuclear energy deserves a place at the table, but should it be an appetizer or a main course? Cost, nuclear waste, and public perception problems will continue—guaranteed.

Energy efficiency and conservation are the "un-fuel" that provides low-hanging fruit for energy savings, but these can only help bridge the gap to an alternative, sustainable energy supply. With the tremendous growth in global human population and the blessings of a growing middle class in developing countries such as China and India, efficiency improvements and conservation can only do so much.

Natural gas provides the cleanest burning fossil fuel, and has grown in popularity, but it shares many of the same location, supply, and price volatility issues as its big brother, crude oil.

Reducing petroleum crude usage remains the biggest and most dangerous global challenge. Petroleum-based fuels make up over 95% of world transportation fuels, and that lack of diversity endangers national security and the world economy. Radical Islam and socialist dictators control many of the world's oil-exporting countries. Islamic lands contain two-thirds of the easy-to-produce conventional crude oil supplies. Christians face even greater perils because dependence on things like oil hides a slippery slope to greed, sin, and idolatry. Islam means "submission to Allah's will," which is not the same as our Christian belief that Jesus is true God and true man, who died and rose again to redeem us from our sins, once and for all. Christians should not submit themselves to dependence on oil, period, and much less on oil in "Allah's lands." When we realize that the United States cannot sustain its present lifestyle for long—even with ANWR, deep offshore oil drilling, and unconventional oil tars and shale—we wake up to the fact that our dependence and allegiances are dangerously off base.

So do we wait for big government to save us, or for the economic markets to guide us through the maze of technological answers? Big government rarely runs anything efficiently, and the economic markets react too slowly to steer the huge energy infrastructure proactively to alternative energy technology. The Christian answer to both is an emphatic no! Government and economic markets will serve as the framework if we utilize them properly, but the true answers reside in Christian application of Scripture to the maze of alternative energy technologies.

We start our journey with the education of the common Christian, who is not typically an energy expert or a climate expert. Since the topics of energy and environment each encompass vast amounts of information (and misinformation), I've tried to provide a "middle of the plate" pitch, to build up knowledge and understanding of where we are and where we need to go. To use a baseball analogy, those "umpires" on the left of the spectrum will likely view my pitch as too far inside the plate, and those on the right will call the same pitch too far outside. I'll let you fill in your own joke about umpires needing glasses.

The mission of *An Inconvenient Purpose* is simple. We seek to strengthen our relationship to Christ while adhering to biblical tradition: through godly stewardship and the amazing world of alternative energy, focusing on how we treat God's creation and how we use our God-given energy resources. Christians all too often ignore biblical stewardship commands. *An Inconvenient Purpose* seeks to establish hope based on biblical wisdom and practical (but not easy) answers to this world's energy and environmental problems.

CHAPTER 1

HOUSTON, WE HAVE A PROBLEM

> How wonderful it is that nobody need wait a single moment before starting to improve the world.
>
> —Anne Frank[1]

As my five-year-old son, Garrett, and I drove past a gas station in the summer of 2005, shortly after Hurricane Katrina had devastated the Gulf Coast, my observant son blurted out, "Daddy, why is the price of gas so *high*? It costs three dollars."

I was amazed at my perceptive child; like a sponge, I'm sure he has absorbed my feelings of frustration from a recent fill-up. As I stumbled through a response while nervously passing a cell phone preoccupied SUV driver, I realized the complexity of this simple question. Supply and demand, greedy oil companies, gas-gluttonous SUVs, hurricane-wrecked refineries, secretive oil cartels, Islamic radicals, and unstable Middle East regimes are just a few reasons for the surging gas prices. However, before I could finish an answer to Garrett's first question, he followed up with another gem.

"Daddy, do we *have* to use gas?"

Caught off-guard by the pure and simple logic of this question, I realized how often we adults blindly accept gasoline and fossil fuels as our only real choice. We take it for granted, since that's all we've known. Still a little flustered, a better response luckily came to me. To put it in terms young Garrett could better understand, I slipped into Ward Cleaver mode and explained, "Well, we need to use some kind of fuel, but it doesn't have to be gas. Think about this, Garrett: we need food to fill our tummies, just like we need gas, or some kind of fuel, to fill our cars and trucks, right? Well, just like there are choices of food, there also are choices of fuel. That's why mom and dad want you to eat your vegetables instead of just cookies. It's a better choice. Someday, we need to find a better choice than gas for our cars."

Cookies and Carrots

When Garrett and I got home, I went into the kitchen to prepare my always-hungry boy a snack. I decided to give Garrett a test. I pulled out seven carrots and a cookie and placed the pile of carrots next to the cookie on the counter.

I called Garrett to come have a snack, and said upon his arrival, "Garrett, you can have the cookie or the carrots, but not both. If you chose the cookie, you have to walk around the table seven times before you can eat the cookie. If you choose the carrots, you can go straight for them."

He glanced at the carrots and then he stared at the cookie. He looked up at me, as if waiting for me to tell him that he had better choose the carrots, but I gave no sign. Finally, he ran around the table seven times, and then happily gobbled up the cookie.

I asked him, "Why did you pick the cookie?"

He replied, "I knew it was better to pick the carrots, but the cookie looked so good I couldn't help it, Dad. And it was kind of fun running around the table. Just leave the carrots there, Dad, I'll eat them later."

With that, he scampered off to play. I chuckled as Garrett disappeared into the family room. That stellar parenting moment in the car had not sunk into Garrett's brain as I had thought. The experiment did prove Garrett is more like "the Beaver" than I'm like Ward Cleaver. I was a little disappointed by the results of my experiment, but not surprised. *This is so much like our energy choices,* I thought to myself. *We know what is better for us, but the winding road to decadence is more fun than the abundance and goodness of the straight path that is right there in front of us. We figure we'll just take the straight path tomorrow . . . .*

The world we live in today offers more choices than ever before; there are even fourteen varieties of Oreo cookies on the market. Alternative energy technologies are becoming more abundant and beginning to hit their stride in the mainstream. Future technologies will soon expand those choices even more. Electrical, chemical, and mechanical energy surround us everywhere. I'm not talking about the "May the force be with you" type of energy from the *Star Wars* movies. Energy is simply anything that has inherent power, or the capacity to do work. Without energy, we cannot power a light bulb or a vehicle.

The subtle differences between energy and power often confuse people. The definitions closely relate to each other. Simply put, energy is a measure of potential work capacity, like a bucket of water. Power is a measure of how fast that work capacity is used over time, like pouring the water from the bucket onto a water wheel. The faster the water pours, the greater the power.

One common measure of energy, the calorie, provides us with an example. Let's say I eat an apple that has 100 calories of energy. It doesn't matter if I exercise or sit on the couch, the apple's energy content remains 100 calories. However, I receive more power from the apple if I exercise, since my body uses the apple's calorific energy faster. Simply, power equals energy divided by time. Think of what happens to the apple's 100 calories if I sit on the couch. The calories (energy) have no work to do, so my body stores the energy as fat to burn at a later time—maybe never. That's why it's important to pair diet (the proper type of energy, such as proteins instead of sugars) and regular exercise (the power generated by burning the energy at a faster rate so that it doesn't get stored as fat). Other types of energy and power work much the same, except we use different terms and definitions because we're not talking about the body's use of energy. With the body, we're talking about energy measured in reference to carbohydrates; with the power and energy industry, we're often talking in reference to hydrocarbons; they're two sides of the same coin.

Like everything on earth, energy is a gift from God. Two categories result from this gift from above: *energy sources* and *energy carriers*. Energy sources include crude oil, natural gas, uranium, and the vegetables my son eats to power his body. Energy carriers include electricity and hydrogen (and Garrett's two legs). To keep things simple, the term "energy" will refer to both energy sources and energy carriers.

Why should we discuss energy? What's the purpose? And why now, at such an inconvenient time, when we have so many other things to do with our time, talents, and money? Because we face a perfect storm of many factors, and we need to be proactive in finding the right shelter before we're blown away. Fuel prices, especially the cost of oil, have ridden a rollercoaster of late. Even when oil prices retreat, the underlying supply issues do not disappear, they just simmer until the next crisis.

A moral decline permeates our society as energy-related companies like Enron have rocked the economy with scandals. Environmental impacts of pollution threaten both water and air supplies. We scrutinize the impact of greenhouse gases on global warming, thought by some to have devastating potential to life as we know it. Domestic oil production and refining, hammered by hurricanes Ivan, Katrina, and Rita, showed our vulnerability to nature. Hostilities in the Middle East continually threaten critical oil supplies. Consider also the mind-numbing effects of the financial and stock market meltdowns in late 2008, and the world seems adrift in a stormy sea. A true energy crisis at any time could act like a deadly lightning strike in the perfect storm, immobilizing our economy and our way of life.

Do We Really Have an Energy Crisis?

Those of us old enough to remember the 1970s remember the energy crisis that peaked in 1973 and 1974. We endured long lines at gasoline stations, and experts predicted that oil supplies would run out in ten years. It never happened; the predictions were wrong.

Is today any different? A new breed of experts is reshaping arguments from the '70s to warn of impending doom in energy markets. Other experts claim we have plenty. Who is correct? Even more importantly, why should we care—especially since we seem to have no personal control over the oil supply situation? What do we do about it? Do we sit tight or take drastic measures?

We will find purpose and address these questions from a Christian perspective because neither economics alone, environmentalism alone, government alone, science alone, nor peace in the Middle East alone can solve the puzzle. These pieces must form a pathway. God offers us paths of moral high ground that may be difficult to walk, but are well worth the effort. Just as Garrett decided to take the path to the cookie with plans to go for carrots tomorrow, we, too, need to evaluate our present path of energy choices. Tomorrow fast approaches.

But first, let us assess our current energy situation. Gasoline and diesel fuels, refined from crude oil, are the most common and widely used types of energy. Unfortunately, there aren't many readily-available, cost-effective transportation alternatives to gasoline-powered automobiles and diesel-powered trucks. Because of the almost exclusive dominance of the combustion engine in vehicular transportation, the world appears at the mercy of crude oil and those who have it. Alternative fuels like ethanol and bio-diesel will provide some relief, but many believe they will not provide a 100-percent solution to the problem. Other technologies, such as gasoline-electric hybrid vehicles, are suitable as "transition technology," but only delay the inevitable due to their semi-dependence on gasoline.

Next, let's talk about energy independence. It's a popular term, especially during political election campaigns. Basically, it means not relying on others, especially foreign countries, for energy resources. However, the United States presently imports more than 58% of our daily consumption of oil from foreign countries.[2] And that percentage will likely grow during the next couple of years, barring the occurrence of an undesirable event, like a long-lasting recession. Demand and prices dropped drastically in 2008, but like the stock market, it's only a matter of time until oil prices rebound even higher. Many of these foreign countries have governments at odds with industrialized countries like the United States, or have deep-seated religious or cultural differences with Western society. The United States, no matter right or wrong, is often perceived abroad as the primary enemy. That tallies up only some of the problems with oil and its byproducts.

Similarly, the other form of energy we use every day, electricity, has its own challenges. Generating electricity requires a fuel source. Approximately 90% of the electricity generated in the United States uses oil, natural gas, coal, or uranium as the fuel stock.[3] Each fuel source in this group has limitations. For instance, the price of natural gas can fluctuate drastically because of supply and demand volatility and dependence on foreign sources. Those concerned with air quality and the environment consider coal—the fossil fuel workhorse for electricity generation—a dirty choice. The often-reviled nuclear power plant, the other large provider of electrical power, remains hampered by public concerns of safety—from fears of accidents to fears of terrorist plots.

The generation of electricity is not the only challenge for the electrical system. The electric grid—the transmission and distribution network that brings electrical power to homes and businesses—is aging and requires extensive capital improvements. Without these improvements, blackouts are more likely, as well as increased susceptibility to terrorism.

The rapid growth of computer usage, microwave ovens, and other devices contribute to power quality problems for sensitive equipment connected to the electrical distribution system. Think of it as using a clothesline to dry your fine china. Sure, it works in theory, but the clothesline is not designed for the weight of the china, and the china is more susceptible to breakage. Likewise, blaming the utility company for the computer crash makes as much sense as blaming the clothesline for breaking the plate. It is misguided criticism at best. We'll learn more about power quality in Chapter 4.

We can compare our current energy situation to a blanket. It's as if threads of blessing and curse intertwine to weave the energy blanket, making it impossible to remove one type of thread without unraveling the whole thing.

The fabric fibers of "blessing" represent lack of panic. If we look at oil and electricity globally, it would be easy to come up with many doomsday scenarios and reasons to feel anxious or depressed. However, many people don't have the energy to worry about energy. They don't have the time or feel the need to stop and contemplate the big-picture energy situation. Panic is contained because the majority of people are content or indifferent to energy issues, except for short-term economic worries at the pump and the electric meter. Thankfully, although people often complain, they assume more of it will come from somewhere.

The fabric fibers of "curse" represent apathy and an innocent lack of urgency. People blindly accept the way it is regarding energy issues. It's funny that what triggers attention to the energy "curse" is the same short-term economic worry at the pump and electric meter.

The miniscule difference between the blessing and the curse is, in many ways, the crux of the problem with energy-related issues. Although gasoline and electricity play crucial roles in everyone's daily lives, most people and companies do not think past the short-term financial aspects of energy.

Some feel we have no real energy crisis at all, that we are making a mountain out of a molehill. After all, America and the industrialized world created mighty economies on oil, fossil fuels, and the like. Why not sustain it as long as we can? We like the lifestyle we have and do not want to make any sacrifices to our standard of living. The problem with this kind of thinking is that we cannot sustain our fossil fuel economy. It will run out eventually. Whether it runs out in ten years or two hundred years, it will end.

Does this mean everyone should abandon his or her car and microwave oven tomorrow? No, but changes will occur and must occur, whether we like it or not. The infrastructure to support fossil fuels is impressive and extensive, which also means that overnight change won't happen. It takes time to plan and develop alternative energy means. Moreover, the cost of fossil fuels will continue to climb as supplies dwindle and demand grows. This is pure Economics 101.

In addition, national security, the environment, religion, politics, and terrorism revolve around long-term energy issues. Because of too much short-term economic focus on energy decisions and not enough consideration of other aspects, a less-than-ideal picture exists for today's energy markets. In fact, past energy decisions by companies, governments, and individuals have unintentionally fostered political and economic instability, environmental dilemmas, religious upheaval, global terrorism, and dependence on foreign regimes.

Attempts to sustain cheap energy are rife with potential dangers:

- In order to secure availability of oil, there is great pressure to pander to the ideological and political will of Islamic nations from the Middle East, including some who sponsor terrorism via supporting certain Islamic radical-fundamentalist schools and movements. We presently receive about 16% of our imported oil from the Persian Gulf.[4]

- Oil supplies from other foreign countries can be equally problematic. For instance, dealing with Venezuela's government under the leftist iron grip of Hugo Chavez can be difficult and dangerous. About 10% of our imported oil comes from Venezuela.[5]
- Although approximately 18% of our imported oil comes from our friendly Canadian neighbors to the north,[6] increasingly this oil will come from "oil sands" production, which poses troubling environmental and economic questions.
- Grave danger exists in fighting foreign wars and occupying countries for securing access to oil. The overwhelming human and economic costs of war do not get paid at the pump, but we cannot escape them; we pay these costs indirectly.*
- When fossil fuel costs start to climb out of control, or when the economy falters, it will be tempting for companies to cut corners on expensive pollution controls or lobby intensely to reduce costly regulations designed to protect the environment.
- Nuclear power plants, regardless of their overall safety record, are always just one accident away from public support meltdown.
- Short-term financial gain makes it tempting, wittingly or unwittingly, to participate in deals with unscrupulous people and companies of all sorts. Some cannot resist these temptations to leverage their future on a fulcrum of greed. (Enron, AIG, Bernie Madoff, Fanny Mae, and Lehman Brothers, to name a few).
- Some believe that drilling for natural gas and oil, mining for coal, or locating facilities in environmentally sensitive areas are not valid options because they cause potentially irreversible ramifications to the environment. Others believe it's economic suicide not to explore and develop these natural resources. Deciding whose opinion to trust, and where to draw the line on what's acceptable, is difficult.

---

* For the purpose of this book, I am not expressing my opinion for or against U.S. involvement in Iraq or elsewhere in the Middle East because I do not want to detract from my message that both sides benefit from pursuing alternative energy. I also want to express my sincere gratitude to our U.S. military service men and women and all that they sacrifice for each other and our country. I do not mean to belittle the human cost of war by referring to economic costs. The focus on economics merely makes a world-view point. Economics never has, and never will, accurately measure the value of human life.

To date, conventional wisdom has failed to solve the world's energy problems. Many believe that if free markets existed without government subsidies, tax incentives, environmental regulation, and the like, we would be in a better place. Unfortunately, the past cannot be changed and the future will not be one of unfettered free market capitalism. It just won't happen, nor should it. Once established, the cycle of government subsidies, tax incentives, and the like is very hard to break. The economic bailout and stimulus packages forced upon us by circumstance by the collapse of the economy will usher in a new era of big government spending. Since energy is a worldwide commodity, all the nations of the world would have to stop these government subsidies, incentives, and tariffs simultaneously to have a level economic playing field. As the saying goes, this will happen "when pigs fly."

In other free market arenas, the electrical industry's moves toward deregulation and market-driven consumer choices have not always succeeded as planned. Some states, such as Texas and Pennsylvania, have had success in deregulation efforts; some have had more difficulties. In California, for example, steps toward electric utility deregulation led to rolling blackouts, lost revenue, and a public backlash that played strongly in the removal from office of Governor Gray Davis. Greedy energy traders like Enron manipulated the market profit, completely disregarding public well-being.

On the other end of the spectrum, many believe that more government involvement is the solution. Proponents claim that free markets will not act on their own to protect the environment and that new ecology-minded technologies require developmental support and early market support by government mandate. These claims may very well be true, but is the government the best institution to develop and nurture long-term environmental and alternative energy policy? Governments have a history of inefficiency and poor guidance in many arenas. To rely on them too heavily in dictating energy policy would be a mistake.

The fact is that we will have to deal with the situation as it stands, but also be flexible enough to adapt as the world changes. This means working with governments, both our own and abroad. It means dealing with huge corporations. It means sorting through a maze of energy technology choices. It means finding the right environmental balance. It means finding energy independence without abandoning the third world. It means confronting terrorism. It means sharing finite resources across the globe and utilizing them for the best benefit of all. It means abandoning feelings of entitlement and becoming grateful for the standard of living we enjoy as a country. It means examining what is truly important to our spiritual core.

It's easy to feel overwhelmed by the magnitude of this task. How can any one person make even a small dent in the energy problems we face? The burden has proven too great for world powers and global corporations to handle by themselves, although few would openly admit it.

This is where trust in God is important. To the world at large, and for too many Christians, God may appear to have no part in energy matters. What we perceive as God's detachment is really His patience in waiting for us to turn to Him for direction. We are God's representatives here, in charge of this place. He expects us to take care of it as He would. With the free will God gives us, will we choose cookies or carrots? If my son had asked for my advice in the snack experiment, I would have strongly recommended the carrots. If we prayerfully seek God's guidance on energy matters, the correct answers, whatever they may be, will certainly come with more clarity. Of course, God will not simply make all our difficulties go away. He doesn't usually work that way. But we can receive a new understanding and attitude that puts our hearts and minds at ease. Proverbs 3:5–6 states it beautifully: "Trust in the Lord with all your heart and lean not on your own understanding; in all your ways acknowledge Him and He will make your paths straight." Until we acknowledge that we need help in the situation, no help will come.

CHAPTER 2

A-E-I-O-U:

AMERICA, ENERGY INDEPENDENCE, OPEC, AND YOU

We all know the vowels of the English language: A-E-I-O-U. They also work nicely as a mnemonic device to remember several of the key items you need to know about the petroleum industry:

- A = America, the country with 5%[1] of the world population which consumes 24%[2] of the yearly petroleum resources.
- E-I = Energy Independence, the goal America (and other nations) would like to achieve for national and economic security.
- O = OPEC, the Organization of Petroleum Exporting Countries, a group of countries which controls an estimated 70%[3] of the world's conventional oil reserves.
- U = You, the person affected by petroleum and who affects petroleum usage.

A: America

"America the Beautiful" has become "America in Bondage." Even George W. Bush, who many people accused of being soft on "Big Oil," declared in his 2006 State of the Union address, "We've got a real problem. America is addicted to oil." This is a profound statement coming from a president with roots as a Texas oilman. Nevertheless, it's accurate; America has developed a nasty habit of consuming too much oil. We expect it to be there for us and we expect it to be there *now*. The U.S. consumes 24% of the world's oil while having only 5% of the world's population. That prolific consumption is both a sign of our industrial might and our crippling weakness.

The exposure of our oil addiction highlights what a crutch oil has become to our present and future strength. As a country, we need to decide if we should throw the crutches away or continue to use them to prop ourselves up. But here's the rub—we cannot just declare ourselves healthy and throw the crutches away. We have to rehabilitate and strengthen our legs. To confound the problem, the doctors (energy experts, politicians, business leaders, environmentalists) cannot agree on a rehab program. Therefore, we continue using crutches while our legs continue to atrophy.

The transportation sector of the U.S. economy is the most dependent on oil, as well as the most at risk. With more than 240 million[4] petroleum-driven vehicles already on the road and less than 4%[5] of them running on something other than gasoline or diesel, one can see that we are "over the barrel."

It is important to be realistic. Even if, by some miracle, hydrogen-powered vehicles appeared on the showroom floor today with a comparable price and performance, many Americans would not abandon their existing sport-utility vehicles (SUVs). Beside the typical fear of the unfamiliar, there is another reason: stranded asset cost. Since autos usually depreciate the moment they're driven off the dealer's lot, the vehicle's owner would lose money on the deal if they sold their present vehicle and bought the new hydrogen-powered vehicle. Many would choose to drive their current vehicle until they perceive that they've squeezed their money's worth out of it. Few would make the economic sacrifice for a single benefit like saving the environment.

However, I believe Americans would step up to the economic sacrifice when multiple causes become more evident: helping the environment, slashing dependence on foreign oil, eliminating erratic and high fuel prices, to name a few. Indeed, this movement has already started with the successful advent of one of the few options presently available, the gasoline-battery hybrid vehicle. Most realize that hybrids, as well as other efficiency measures, help reduce our dependence on oil, but they are not the final solution because they still use significant amounts of gasoline. They are a transition technology at best, but serve an important function in priming the public for coming technologies, such as fuel cells. To help decipher our "oil slick" of a problem, we need to revisit how we got to where we are today.

THE BLAME GAME

In an MSNBC poll reported on *Meet the Press* on April 30, 2006, people were asked who they blamed for high oil and gasoline prices. The results relate only to people's perceptions and not fact, but the results of the survey were interesting: 37% blamed oil company greed, 22% blamed oil-producing nations, 15% blamed the president, 8% blamed themselves (the consumer), 6% blamed federal regulations, 4% blamed congress, and 2% blamed auto manufacturers. Let's look at the three largest scapegoats.[6]

The 37% blame assessed to the oil companies has some pitfalls. For one, government has repeatedly called for investigations into oil price gouging since the 1970s, but each time the only thing gouged is the taxpayer's pocketbook for conducting the unfruitful studies. Secondly, profits are huge because the industry is huge, as well as risky. These are basic concepts in a free market system. Often, the oil company's earnings per share are less than other industries that do not receive the public wrath of price gouging claims. If McDonalds has an exceptional financial quarter, the government does not call for investigations while the public demands price reductions on Big Macs and Quarter Pounders. Big oil companies are not saints, but neither are they the complete villains the public perceives them to be. In fact, their financial strength and infrastructure assets might be a key factor in righting the ship. For instance, existing gas stations present opportunities for retrofitting to new alternative energy fueling systems, such as compressed hydrogen. The facility could then serve both types of fueling needs, gasoline and hydrogen. It's a better solution than starting from scratch.

Likewise, for the oil-producing foreign nations, the public exaggerates their culpability. They possess a natural resource that trades on the world market, based on the economic laws of supply and demand. The free trade economy is not perfect and gets manipulated sometimes, but all-in-all, they have a right to expect top dollar for their resources. If the situation was reversed and the U.S. controlled vast reserves of oil, I'm sure we would respond in similar fashion. The price run-up to $147 a barrel in July of 2008 reflected more influence from market dynamics than it did from manipulation by OPEC nations. Not to defend the self-centered actions of OPEC, but since OPEC's formation more than forty years ago, people have consistently found it a convenient scapegoat for high oil prices. We forget that our addictive dependence to oil is what gives OPEC and other oil-exporting nations that power over us.

Regarding our third scapegoat, sitting presidents, whether democrat or republican, have little control over short-term oil prices; the market drives oil prices. There is very little they can do to relieve prices without creating worse dilemmas. We already pay much lower taxes than most of the industrialized nations, especially those in Europe, and forgoing the taxes would negatively affect roads and bridges, both construction and maintenance work. The strategic petroleum reserves could only supply the country for a few months and are not intended to relieve financial burden, but to cushion against natural disasters such as hurricanes, and provide a temporary national security cushion for global supply disruptions.

We added another evil villain to the list in 2008. Oil speculators received much public blame for the petroleum price roller coaster. Many said investments flowed in and out of the oil market which had nothing to do with supply and demand, thereby corrupting the price. In a *New York Times* op-ed column, others, such as Nobel Prize-winning economist Paul Krugman, claimed differently. He surmised that true oil production restrictions and strong demand growth in China and other emerging economies justified the inflated price.[7] "Krugman also pointed out that the most insistent claims of an oil bubble were coming from conservatives with an interest in denying the reality of peak oil. 'Traditionally, denunciations of speculators come from the left of the political spectrum. In the case of oil prices, however, the most vociferous proponents of the view that it's all the speculators' fault have been conservatives—people whom you wouldn't normally expect to see warning about the nefarious activities of investment banks and hedge funds. The explanation of this seeming paradox is that wishful thinking has trumped pro-market ideology.'"[8]

Still, after the crash in oil prices in late 2008, one cannot deny speculators had more extensive influence than first thought. However, the crude oil price undershoot seems as strong as the overshoot in the summer of 2008. It took the worst economic fallout since the Great Depression to cause the price of oil to collapse. The physical limits of the amount of oil in the ground remained unchanged for both price swings, except for the oil we continue to pump from the ground.

GAS GUZZLERS

Part of the problem is what we choose to drive. When gas prices sank in the late 1980s and through the 1990s, Americans changed in droves to SUVs and light-duty trucks. With low prices at the pump came low importance placed on fuel economy. In fact, the average fuel economy was better in the late 1980s than it was in the early 2000s. Instead of focusing on fuel economy, engineering efforts went into increasing the power and size of vehicles. It would be easy to fault the automotive industry for the dilemma we are in, but truthfully, there is plenty of blame to go around. Auto manufacturers simply build what Americans want. Now Americans want something else and Detroit has failed to anticipate those changes. Does that mean we should all ride around on mopeds? No, but it does mean we should stop grumbling and focus our energy on defeating our oil addiction, rather than assessing blame.

There is nothing morally wrong with the desire for more power and performance out of a vehicle; just be willing and prepared to pay for it in the short term. Sometimes, special circumstances—like large families, for example—make those large SUVs and minivans perfectly justifiable. It's better to use one vehicle with poor mileage than to double up on trips with smaller, more fuel-efficient vehicles. Personally, I think we'll see the day when renewable hydrogen-powered fuel cell vehicles with high-torque DC motors at each wheel (cranking out unbelievable power) will outperform today's internal combustion engine, at less cost and with zero pollution emissions.

With regard to efficiency, much political debate has occurred regarding raising the Corporate Average Fuel Economy (CAFE) standards. In 1990, the CAFE standard for passenger cars parked at 27.5 mpg and remained there for years.[9] Eventually, politicians acted by putting the

Energy Independence and Security Act of 2007 on the books, which requires, in part, that automakers boost fleet-wide gas mileage to 35 mpg by the year 2020.[10]

Some environmental groups believe that current fuel economy mandates are too modest and phased in over too many years to have much effect. Others say it's more practical to implement fuel efficiency increases in moderation. Still others say CAFE standards have been ineffective at curbing our fuel usage, possibly even backfiring. Regardless of who is correct, the historical fact remains that when gasoline prices turned downward in the mid 1980s, lawmakers turned their attention away from fuel efficiency directives. This illustrates the rebound effect of efficiency measures—when prices fall, the original impetus disappears and the will to enact efficiency improvements wanes.

However, we shouldn't focus too much on the band-aid of efficiency improvements. Raising the CAFE standards simply forces efficiency improvements on automotive companies which is, at best, a short-term fix. Our efforts are better spent searching for a long-term solution which involves a carrot, rather than a stick approach.

Americans must first realize that we are not alone. Out of our sight, and half the world away, live over 2 *billion* Chinese and Indian citizens. Developing nations such as China and India will dominate energy demand in the future, just as much as America does today. As their appetite for oil increases, our impact on world prices will decrease. The consequences are clear: where America could once control her own destiny concerning oil prices by cutting consumption when necessary, the world's growing need for oil is diminishing—or eliminating—that option for the United States. We must now concede that our economic superiority can no longer depend on cheap oil.

E-I: Energy Independence

Election-hungry politicians talk a lot about energy independence. Unfortunately, most politicians never get beyond campaign rhetoric. The term *energy independence*, simply put, means not needing foreign resources to supply the energy needs of a country. Critical energy fuel and infrastructure changes demand long-term commitments and painful price tags—two items politicians avoid until forced into it. To do otherwise would be political suicide, like a turtle crossing a busy highway.

We must address our dependence on foreign oil. Western nations have gone to war in the Middle East and propped up horrible dictators to secure access to oil. As one of former President Reagan's aides, Lawrence Korb, bluntly admitted concerning America's involvement at the start of Operation Desert Storm, "If Kuwait grew carrots, we wouldn't give a damn."[11] As a patriot and a Christian, it's hard to admit the truth of that statement, but the fact remains that America's political policy often ignores the actions of cruel dictators and corrupt governments all over the world. Not because we are afraid of them; sometimes it's because those countries have paltry oil or natural gas reserves, or other strategic value. Other times, it's because they are a "friend" who has oil reserves. Case in point: Saddam Hussein was once a "friend" of the U.S. when he supplied us with oil and served the purpose of counter-balancing Islamic fundamentalism in Iran. Nevertheless, when he threatened the balance of power in the Middle East by invading oil-rich Kuwait, Saddam's sins were no longer ignored. The former "friend" became an enemy, although he was the same cruel dictator all along.

One could make a strong case that every king from Saudi Arabia since the 1940s falls in the "friend" category because of oil resources. This regime remains extremely repressive, especially to women, but still we sell billions of dollars in military technology to them and purchase tens of billions of dollars of oil from them.

Likewise, when a totalitarian country lacks oil reserves, we often do not take action with the same gusto we showed when neutralizing Saddam Hussein. How many in the West can name the cruel dictator that rules Burma (Myanmar)? Or how about naming the dictator of Sudan, who has been in power since 1989? Both of these "gentlemen," Than Shwe and Omar Al-Bashir respectively, were in *Parade* magazine's article, "The World's 10 Worst Dictators"; in fact, they placed in the top five in 2008 and in earlier years also.[12] The list for 2009 still has them in the top five. Unfortunately, human nature leads us to care only when it affects us directly. That is how oil corrupts, why there will never be peace in the Middle East if oil is a primary fuel, and why striving for energy independence is so important.

Understanding energy independence requires understanding the nature of oil reserve impacts on national security. Since our own reserves of oil reached a production peak in the early 1970s, imported oil has increased in percentage and correspondingly weakened our national security, both economically and militarily. The transfer of wealth to oil-rich nations tears at the very fabric of our nation. Foreign oil dependency is more than an economic nuisance; it holds the seeds of national disaster.

One method of economic and political forecasting is scenario planning.[13] Shell Oil has successfully used this method to forecast and plan future business moves, such as positioning themselves in natural gas markets before the Soviet Union fell. Scenario planning could help us understand the dangers of energy dependence on foreign oil.

Take, for instance, the potential impact of a major terrorist strike on Saudi Arabia's oil infrastructure, or the fall of their government to radical Muslims who refuse to sell oil to Western infidels. Over 10 million barrels of oil per day would be removed from the market.[14] The ability of other oil-producing nations and companies to pick up the slack would be severely taxed or impossible. Approximately 20% of the world's known conventional oil reserves would be in jeopardy.[15]

But how do we move past political rhetoric and move towards energy independence? The key lies in how we use all of our God-given resources, including oil, and taking care not to let them become idols. Anything that comes between God and us can be considered an idol, including oil. Energy independence has everything to do with energy dependence on God. Biblical wisdom teaches us to trust first and completely in the Lord, and God will provide. That doesn't mean we are to do nothing. Instead, we seek energy independence from other resources God provides. Oil is not sustainable or under our control; it is an energy dependence ball and chain.

O: OPEC

What is the difference between the Clampetts and the Saudi Royal family? One used to be poor as dirt, and the other used to be poor as sand, until oil made them rich. Bad joke aside, I realize it is unrealistic to compare a fictitious family from a 1960s TV series, *The Beverly Hillbillies*, to the Saudi royal family. I mean no disrespect to the royal family—or to Hollywood. Nevertheless, families and whole nations who held nothing in material wealth have become, in just a few generations, world leaders and endowed nations. The vehicle for that transformation was petroleum, and OPEC is the historical legacy that followed.

In the early years of oil discovery in the Middle East, around the 1920s and 1930s, the kingdoms of the region were devastatingly poor. They were remnants from the fallen Ottoman Empire, the last Muslim empire that united the Middle East. Most of the countries which later proved to possess oil were part of the British colonial empire. This includes Saudi Arabia, a consolidation of warring tribal realms brought under control by a conquering leader, Abdul Aziz ibn Saud.

At first, Middle Eastern countries did not produce the petroleum from the oil reserves themselves; they sold the concessions (or rights) to search for the oil in order to raise money quickly for their financially strapped regimes. Oil companies bought the rights to explore, drill, and produce the oil in a predetermined tract of land, and paid royalties to the government on the amount produced from the wells.

Saudi Arabia made their concessions to an American company, Standard Oil of California (SOCAL), in 1933.[16] In those days, the Great Depression engulfed the world economy, including an oil glut where crude oil sold for 10 cents a barrel. The Western world's appetite for oil did not yet exist. It remains a miracle that the concessions deal with the American company SOCAL ever took place, as quoted from Matthew R. Simmons's book, *Twilight in the Desert:*

> After Abdul Aziz read the final draft agreement, he turned to his finance minister, Adullah Sulaiman, and said, "Put your trust in God and sign." As it turned out, God was clearly looking out not just for Abdul Aziz and his subjects. God also must have been looking out for all of the advanced nations of the world, as it would take only another two decades for these nations to develop a ravenous appetite and critical need for the oil that lay beneath the sands of the SOCAL concession.[17]

The mother of all oil fields, the prolific, oil-producing, super-giant named Ghawar, consequently came into existence in the following decades. Although the financial and geopolitical value of these concessions remained unknown at first, other countries also tried to acquire these oil rights or schemed to take them away from the Americans. Most notable of these attempts included the British, who dominated the politics of the region, and the Germans under a rising leader named Adolph Hitler.

In Hitler, the Saudi king found a kindred spirit because of their mutual disdain and abhorrence for the Jews. In spite of this, the Saudi ruler could not bring himself to trust Hitler to provide the financial support Saudi Arabia desperately needed. Ibn Saud's reliance on the financial support of the Americans and British kept Saudi Arabia in the fold.[18] By the grace of God, these oil rights did not fall under Hitler's control.*

The Americans received the oil concession in part because they were not the British, since the region had long grown tired of British colonial rule. If not for these early concessions, the United States would probably not have become the premier world power and industrial leader it is today. On the other hand, we would also not be as far out on the limb of oil dependence.

After World War II, in sympathy to the Jewish plight, a homeland in the Hebrew Promised Land arose by United Nations (UN) mandate. The formation of the nation of Israel set off a powder keg of protest in the Arab world, which helped precipitate the nationalization of Middle East oil fields and the formation of the OPEC cartel in 1961. The stated purpose of OPEC is to negotiate oil production prices and future oil concessions with oil companies.

Many assume OPEC is purely a Middle Eastern cartel, when in fact OPEC has members from South America, Africa and, until recently, Southeast Asia. In fact, Venezuela was one of the five founding members. Still, there is a very distinct Middle Eastern flavor to the group, whose current members in 2009 include Algeria, Angola, Ecuador, Iran, Iraq, Kuwait, Libya, Nigeria, Qatar, Saudi Arabia, United Arab Emirates (U.A.E.), and Venezuela.

---

* Lack of German petroleum access played a main role in the failure of the Nazi army. In fact, Germany's General Rommel was poised to overrun Africa and the Middle East, but his effort stalled due to lack of fuel. If he had succeeded in taking the Saudi oil fields, the outcome of WWII may have been considerably different, or the war prolonged.[19]

The percentage of the OPEC petroleum reserves contributed from the Middle East is approximately 80%, and the twelve countries that make up OPEC overall provide an estimated 70% of the world's estimated recoverable oil reserves.[20] The exact percentage of the reserves remains veiled in secrecy since OPEC does not allow third-party verification of their oil reserves data. Regardless, the volume of oil is substantial. Many doubt the truthfulness of oil reserves data, and the uncertainty adds an element of instability to an already unstable situation. The worry centers on the lack of a downside to inflating oil reserve numbers by OPEC nations, since verification does not exist and production quotas depend on reserves. The doubt of oil reserves data between countries within OPEC even exists. Matthew Simmons speaks of the secrecy pitfalls in his book, *Twilight in the Desert:*

> By veiling its oil operations in secrecy and refusing to provide credible data to support its claims about reserves, production rates, and costs, Saudi Arabia (and other producers) served its customers, the consuming nations, poorly. Energy planners the world over were forced to base their calculations on *assumptions* rather than verifiable information, a circumstance that has undoubtedly had harmful consequences for all energy stakeholders, producers and consumers alike.[21]

Not only is the exact quantity of available oil a concern, but also the location of it.

The geopolitical volatility of the Middle East region emphasizes the potential for disaster. One colossal terrorist strike or regional government military action at the Strait of Hormuz could cripple approximately 40% of all the world's oil trade and 80% of the oil exported from the Persian Gulf region.[22] Realizing that the majority of OPEC's major players surround this region creates cause for concern, since daily oil production and the bulk of the world's proven oil reserves remain clustered in such a small region. So many items remain beyond our control in the U.S., such as OPEC itself, conflicting political interests with the Middle East, terrorism, and radical Islam.

Disturbingly, the nations that comprise OPEC are either predominantly Muslim states, or hostile to the United States, or both. The fact that we have been in this position for so long has numbed us to the seriousness of the situation. We should pray every day that we can avoid the tipping point until our dependence on oil is defeated. This does not suggest forbidding crude oil purchases from Islamic nations; it means we should not depend on any country, region, or resource so completely that we cannot judge what is right and godly. The temptation to pander to foreign nations who have vast quantities of crude oil reserves endangers us economically, politically, morally, and spiritually.

From the Middle Eastern perspective, oil serves as the catalyst for upheaval in economics, politics, and religion, which has given rise to pan-Arabism, OPEC, nationalization movements, Palestinian-Israeli conflicts, and radical Islam, although they view these events very differently than Western industrialized nations. One could claim economic reasons for OPEC's formation, versus political or religious reasons. I consider those claims shortsighted, because industrialized nations often underestimate the impact of religion on political and economic life in Muslim nations.

You can see one of the first great examples of this in the 1973 OPEC-generated oil crisis, created by a perfect storm of economic, religious, and political components. Religion certainly played a role, as the United States' backing of Israel in the 1973 Middle East war gave impetus to the oil crisis. Saudi Arabia held deep resentment for the United States, dating back to the establishment of Israel as a nation in 1948. The resentment smoldered continuously as America consistently sided with Israel over Palestinian and Arab interests in the region. The seeds of the formation of OPEC sprang up in the soil of a Jewish nation in what Islamists considered Arab-Palestinian lands. The Jewish, Christian, and Islamic religions each lay claim to the region and consider the others' claims false. The religious arguments have gone on for centuries and are unlikely to be resolved anytime soon, barring divine intervention.*

---

* Christians hold just as much claim to the historic significance of the region as other religions, but Christians are the most absent from the region. Why? Some think this means the Christian claim is the weakest, which is not the case. For Christians, Christ's resurrection transcends the material world's importance of a tract of land. Christians stand up for Jesus, the risen Lord, looking up to the heavens, not down at the tomb he resided in for only three days. Although parts of the Middle East are holy ground for Christians as well as Jews and Muslims, Christians can move on because Christ Jesus lives! The focus on God's message (and not the land) took a while to learn, as the Crusades of the Middle Ages prove, but it's true, nonetheless, of the importance of the spiritual over the material.

The strength of OPEC remained unproven and untested until the 1973 oil embargo, when King Faisal of Saudi Arabia used his "oil sword" to punish the U.S. for supporting Israel. The Palestinians had just lost an embarrassing war to the Israeli army. Arab pride was hurt, and losing to the despised Jewish nation made their loss even worse in the eyes of this disproportionately Muslim region. They cut off the oil supply, using their only means to lash out at Israel and their customers in the Western world.

Why was the 1973 oil embargo successful in crushing the U.S. economy? Although America had been the most prolific oil-producing nation on the planet, by 1973 the U.S. imported a substantial amount of oil from OPEC nations. We can drill deeper for an explanation. In 1956, a geological scientist named M. King Hubbert at Shell Oil made a startling discovery. Simply put, he explained the relationship between oil production and time would fit a bell-shaped curve. Hubbert predicted that oil production would decline in the lower 48 states by the early 1970s. Many of his colleagues scoffed at his prediction at the time, but it came true in 1970 when U.S. domestic oil production peaked. Correspondingly, the U.S. economy slipped tragically in 1973–74 in part because the U.S. could not produce the oil required by the country internally when OPEC imposed its embargo. As predicted by Hubbert, crude oil prices became erratic on the backside of the oil production curve, when producible supply couldn't keep pace with demand.

Many experts today have attempted to use Hubbert's methods to predict the peak of world oil production. Although we cannot determine the peak with certainty until after it's past, some experts speculate we've already hit the world oil peak at 87 million barrels a day.[23] Others say (and I tend to agree) that the peak daily volume might top out closer to 100 million barrels a day. Regardless of the exact peak, the danger occurs on the down side of the curve if no alternative fuels arise to fill the shortfall. As proven by the American peak production reached in 1970, followed by the 1973 OPEC embargo, prices can violently spike when production doesn't keep up with demand. The same will be true soon after world oil production peaks, although the peak looks typically more like a plateau that may last a couple of years.

The OPEC 1973 embargo also succeeded in ravaging the United States' economy because the OPEC nations stuck together and held to

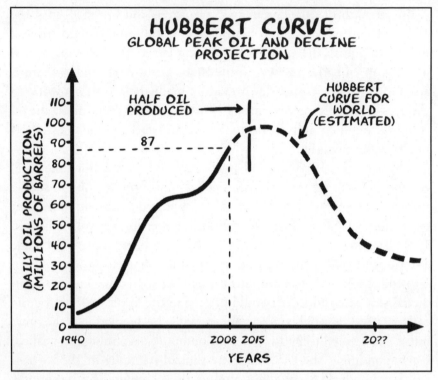

Figure 2.1—Hubbert Bell Curve Estimate for World Oil Reserves.

their declaration. It was the first true test of OPEC resolve and solidarity. They forced the major oil companies and nations around the world to take notice and recognize the power OPEC wielded with its "oil sword."

Since then, the power of OPEC has ebbed and waned, but the member countries retain their bond and influence. The economic importance of OPEC, and oil in general, is highlighted by the timing of economic downturns triggered by sharp increases in oil prices, or danger to Middle East oil supplies:

1973 Oil Embargo
1978 Iranian Islamic Fundamentalist Revolution
1990 Operation Desert Storm (1st Iraq War)
2001 September 11th
2003 Operation Enduring Freedom (2nd Iraq War)

Numerous other disruptions can occur in the world economy based on oil production events or even threats of oil-flow stoppage. These colossal, volatile economic impacts lie mainly out of our control. To make matters worse, our complete dependence on oil makes it impossible to adjust quickly, since we have no large-scale alternatives to implement immediately other than conservation.

Lastly, OPEC also resulted from political stimulus. OPEC has origins in Pan-Arabism, a pro-Arab nationalism movement. Pan-Arabism's roots grew from secular origins, but it's often confused with pan-Islamic movements, especially in Western culture. In the 1950s, Gamal Abdel Nasser of Egypt was one of the first to call Arabs to unite, and of course, the control of oil and the Palestinian situation were central to this pan-Arab movement. The most familiar result of the pan-Arab movement to Western nations is the Ba'ath political party, which came to power in Syria and Iraq decades ago.

Saddam Hussein rose to power in the Ba'ath party. Saddam encompassed most of America's knowledge of the Ba'ath party. Interestingly, the phrase "Ba'ath party" in Arabic language means "resurrection party." What a contrast to the resurrection Christians know in Christ.

U: You

As former President Bush claimed, "America is addicted to oil." How we break that addiction depends on U (you). As with any addiction, most addicts suffer from denial, and breaking the addiction can be painful. Ask any person who has conquered an addiction and they will tell you that rationality goes out the window. The addictive substance—alcohol, drugs, food, pornography, money, you name it—becomes irrationally all-consuming in their life. Some simply cannot stop their destructive behavior. Many don't even want to stop. Or even care.

God commands Christians not to have any false gods. It is the first commandment, "You shall have no other gods before me. You shall not make for yourself an idol in the form of anything in heaven above or on the earth beneath or in the waters below" (Exod. 20:3–4). An idol is anything we place in higher importance than God. Oil can be an idol, if we let it. In fact, it can be an exceptionally dangerous one, since we normally do not think of it as an idol. Couple the idol status with the fact that the Christian world has direct access to only about a third of the world's conventional cheap oil, with the other two-thirds controlled by Islamic countries, and the future looks bleak. But decisions you make can start to turn that around.

Here is one fictional example of oil addiction. I still remember watching the antics of J.R. Ewing on the television show *Dallas*, one of the first prime-time soap operas. The Ewings had oil, money, power, and beautiful people around them, but they weren't really happy. Their life was one struggle after another. We could relate as an audience because we understood the irrational exuberance of oil and power. If the Ewing family controlled a prune juice empire, the effect would not have been the same.

Dallas was a fictitious TV show, but in many ways it represents our infatuation with oil. In a little more than one hundred and fifty years, we've gone from horses to Hummers. Our means of transportation depends so much on oil that we can no longer imagine life without it. In the coming chapters, I hope to encourage you to imagine a different life. Not a life of restriction tied to gasoline rationing, or personal sacrifice due to dwindling oil supplies, but a life with more freedom, personal satisfaction, and a stronger society apart from our oil "ball and chain."

Realizing our dependence on oil is the first step in solving the problem of our oil addiction. Next, we need to do something about it.

Sometimes Y: Why

Sometimes we use the letter Y as a vowel. We should ask our government, our corporations, and ourselves, "Why oil?" Honest and thoughtful answers to this question can lead to surprising insights.

The government might answer like this: because our national security, our future as a world power, and our future as the leading economy in the world depends on the availability of cheap oil. There is nothing wrong with the desire to succeed, as long as we don't sell our souls to do it. All Americans truly hope that the United States remains a successful country. But let's be honest about the real costs of our oil dependence. When we consider the effects of oil on the environment, on wars and geo-political instability, and on oil money trickling down to Islamic terrorists, it becomes obvious that gas costs much more than the price per gallon on the sign.

Regrettably, oil is a limited resource. More regrettably, the conventional proven reserves exist largely outside the United States, and beyond our control. Any change from gasoline or diesel to anything else like bio-fuels or hydrogen will be time-consuming and expensive from a business-as-usual standpoint, but it is necessary. These kinds of fuel and energy infrastructure changes do not happen overnight. Government must help support this change to ensure a vibrant future for our country and the world.

Oil carries a moral cost for us, and for corporations, too.

Corporations and oil companies might answer the question, "Why oil?" in similar fashion to the government, but put it more bluntly: "Because it's cheap and available." Companies resist changes to their industry as much as governments balk at changes to their power base. By nature, corporations enslave themselves to the economic bottom line much more than governments. They have corporate boards, stockholders, employees, and other stakeholders to answer to or they won't stay in business for long, at least not without receiving a government bailout. The risk-adverse answer is often to stay the course.

Typically, people are more apathetic to government fiscal responsibility and don't expect as much in general economic proficiency compared to company performance expectations. Therefore, many people allow government more free reign on items such as new technology spending. Nevertheless, companies desperately need to consider the long-term effects of their business practices, or today's small gains could become tomorrow's huge losses. The bottom line—companies remain as shortsighted as governments to the dangers of our petroleum-based economy.

Many things affect crude oil prices and cause price volatility. This includes supply and demand, oil market speculation, refinery capacity, oil extraction and shipping costs, geo-political conflicts, and natural disasters. The cheap oil that built our world economy and propelled America to the forefront is eroding from beneath us. The real question the corporations should ask is, "How quickly will it deteriorate?"

Figure 2.2—Problems with Oil—Oil Rig Platform Weak Pillars.

The personal response to the question, "Why oil?" may appear obvious: "Because it's there and we don't really have much choice." While
true to an extent, markets do respond to consumer choice. When
consumers purchased SUVs in record numbers, automakers happily
obliged. When skyrocketing fuel costs increased demand for hybrid
vehicles, auto manufacturers started to build them. Consumers have
become willing to pay a premium for a product like a hybrid car, even
when the economic break-even point of the extra cost of the vehicle
versus the fuel savings is years away. Manufacturers take notice and
willingly make the vehicle, even if it's more technically difficult to
engineer and build.

As seen by the government rescue of the "big three" American car
manufacturers in 2008–09, being shortsighted regarding resource

and society changes holds dire consequences. Toyota and Honda have proven more proactive in adapting to a future with oil resource constraints. Toyota leads in battery-powered technology and Honda is strongly pursuing fuel cells as the answer.

Whatever head start other manufacturers may have on commitment to change, American companies must close that gap. The only other alternative is failure. Why oil? For too long the automotive industry smugly resisted and effectively claimed, "Because we're too big to change." Now we see that's not correct.

CHAPTER 3

AN I-C RELATIONSHIP:
ISLAM AND CHRISTIANITY

> I too was convinced that I ought to do all that was possible
> to oppose the name of Jesus of Nazareth. And that is just
> what I did in Jerusalem. On the authority of the chief priests
> I put many of the saints in prison, and when they were put
> to death, I cast my vote against them.
> —Paul speaking to King Agrippa,
> Acts 26:9–10

Do you know the biblical story of how a man named Saul converted to Christianity and became a man named Paul? It is a riveting tale, no matter your religious persuasion. Saul zealously pursued what he was convinced was right and true—only to discover at a life-defining moment that he had it all wrong.

In the story of Paul's conversion, a Jewish rabbi named Saul avidly persecuted the early followers of Jesus Christ, wholeheartedly convinced he was doing the Lord's work. He was convinced he had the moral obligation to do so. Then something happened. A great light from above blinded Saul and struck him down on the road to the city of Damascus.

From that life-changing moment, Saul changed his name to Paul to represent his new calling, turned his life around 180 degrees, and became a zealous follower of Jesus and the leader of people he originally sought to obliterate.

God's plan for Saul transformed him from the inside out and from the outside in. What a testament to the awesome power of God! Saul's beliefs, radically bent in one direction, could go through a metamorphosis before the eyes of Hebrews, Christians, and Gentiles alike. What a testament to Paul's newborn faith, risking the persecution and death which early Christians often faced.

One of the aspects of Paul's conversion, the transformation from persecutor to persecuted, plays out on a larger scale today between Christians and Muslims. Both sides claim persecution by the other. For example, some Muslims claim persecution by Western society (i.e. Christian society in their mentality) because of our greed for their oil. Christians (who make up a dominant portion of Western society) proclaim persecution by Middle-Eastern terrorists (i.e. Islamic society in some of our minds). You cannot deny religion plays a role in the politics of oil.

But deep theological debate over which is the religion of truth is not the point here. Neither is antagonizing or vilifying either religion. This chapter does not seek to make converts, nor does it seek to shrink in the presence of divisive issues. The point is that Saul (Paul) embodied both persecutor and persecuted. Likewise, the two largest worldwide religions, Christianity and Islam, embody both persecutor and persecuted when both sides are honestly considered.

Much of the hostility between Christians and Muslims stems from persecution and the resulting cries for judgment or revenge that follow. There are definite cases of persecution: Muslim faith inflicted upon Christians, and Christian faith imposed upon Muslims. Cases of perceived persecution go both directions. Justice is an important aspect of both religions and both expect justice to result from their persecution. However, Christianity and Islam both receive criticism for being too judgmental. Only God can sit in judgment; we can only judge actions, not motives. Jesus declared in Matthew 7:1–2: "Do not judge, or you too will be judged. For in the same way you judge others, you will be judged, and with the measure you use, it will be measured to you."

Christians and Muslims judge each other's actions, but we also go too far and judge each other's motives, too. The present dilemma in relation to oil causes much distrust of motives on both sides. The perceptions of Christians and Muslims critically differ on many items, including their worldview of oil, and the role of religion in public life, but we also have common ground. Neither side is exempt from the need for forgiveness.

The Spiritual Warfare of Oil

It is always a good idea to get to know your neighbors, whether they live two doors down or oceans apart. For instance, many Westerners persistently assume that most Muslims are Arabic. Only about 18% of Muslims are Arabic.[1] Additionally, many mistakenly believe that all Middle East countries with big oil reserves are equally "Arab." For instance, most Iranians are "Aryan" (or Persian, in Western terms), and not Arab, a cultural distinction lost to many in the Western Hemisphere.

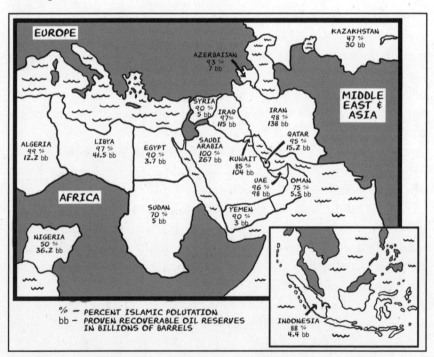

Figure 3.1—Map-Percent Islamic Population vs. Oil Reserves.

An important common denominator which does not receive enough attention in the politically correct (tolerance happy) Western world is that countries with large oil reserves are primarily Islamic.* Many use terms like Persian Gulf oil or Middle East oil, but in essence, it is all oil under the control of Islamic regimes. Is that significant if you consider yourself Christian? As Christians, perhaps we should start asking ourselves questions like, "What affect does my dependence on oil have on my faith? Is my Christian faith compromised by oil dependence on Islamic nations? Is Islamic practice compromised by dependence on Western money?" As you can absorb from the map on page 32, countries with large Islamic populations represent a majority portion of the world's petroleum reserves, estimated as high as two-thirds. As a Christian, I consider this significant, since it sets us up for making bad choices. This dependence, if unchanged, coupled with the downside of eventual peak oil production, could result in the most serious spiritual, economic, and global conflict in modern times.

Saudi Arabia, the birthplace of Muhammad and home of Islam's most sacred sites, Mecca and Medina, has over 20% of the known conventional petroleum reserves, with an estimated capacity of 266.7 billion barrels of crude oil. Iran has 138.4 billion barrels. Iraq has 115 billion barrels. Kuwait possesses 104 billion barrels. The United Arab Emirates reports an estimate of 97.8 billion barrels. The totals for the other top-tier Islamic countries—Qatar, Yemen, Indonesia, Algeria, Sudan, Egypt, Azerbaijan, Oman, Syria, Kazakhstan (about half Islamic), Nigeria (about half Islamic) and Libya—equals 161 billion barrels. A conservative estimate for the world total from the Energy Information Administration (EIA) is 1.33 trillion barrels.[2]

The amount of oil in Islamic lands equals about 840 billion barrels total or about 64% of the EIA estimated total. A barrel equals approximately 42 gallons. One trillion barrels sounds like a tremendous amount of oil. However when you consider the world consumes an average of 85 million barrels a day,[3] the world's estimated supply is not sustainable past 43 years at the 85 million barrels a day consumption rate. Many doubt reserves growth and improved oil recovery technology will extend that much. By the time my son is my current

---

* The term *Muslim* is also used quite often when speaking about the Islamic religion. *Islamic* is the more correct term when referring to the religion, meaning "one who submits to God," whereas *Muslim* means "follower of the prophet Muhammad in the ways of Islam."

age in 2041, the age of oil will be in serious decline. Coupled with the "Islamification" of oil, Christianity (as well as the rest of the world) must soon confront issues we largely ignore today.

Another sign of spiritual struggle: in the Middle East, Christian population has been waning. Also, no Middle Eastern countries with oil reserves have a strong Christian presence. None of this should surprise us, but seeing the numbers in the map above drives home the significance of future energy challenges. The answers to these challenges will hopefully bring energy independence and increased national security so that the West will not compromise its beliefs at the expense of what some would call "economic prostitution." Lebanon is the only Arab country with a significant amount of Christian Arabs (39% of the country's population),[4] but they have no oil reserves or real political power. In the Holy Land, which has no oil and is split between Israeli and Palestinian lands, the instability and persecution of Christians has caused a mass exodus. The Christian population is less than 2%, down from nearly 20% of the population nearly a century ago.[5] According to Reuters, "throughout the Middle East, Christian scholars say, tension is rising between Arab Christians and their Muslim neighbors, who see Christians as belonging to a Western world they blame for the conflict in Iraq and other regional troubles."[6] What does this mean? Christianity and Islam are becoming more polarized in a critical area of the world, crucial in historical significance to the Abrahamic faiths (Judeo, Christian, and Islamic) as well as to the national and economic security of the West via access to oil reserves.

In the last chapter, we talked about OPEC (via Saudi Arabia) using the "Oil Sword" embargo to punish the West for supporting Israel. Since then, other oil-related events have triggered severe economic downturns and recessions. Saudi Arabia deserves some credit for stabilizing oil prices over the last 30 years. But their capability as a swing producer is weakened by the ever-growing demand as we stretch toward the oil peak. Moreover, the Saudi kings and princes may not always be able to rescue oil production. Many say that oil reserves outside of the OPEC/Middle East/Islamic countries are approaching the oil production peak sooner, thereby placing more strain and importance on the "Islamic" supplies. Promising new exploration areas near the Caspian Sea have not lived up to expectations to date, but they too would be under largely Islamic territory.

And, as Jeremy Rifkin writes in his book *The Hydrogen Economy*, "Oil is increasingly viewed by a younger generation of Muslims as the 'great equalizer,' a spiritual as well as geopolitical weapon that, if Islamized in the service of Allah, could lead to the second coming of Islam. King Fahd of Saudi Arabia sensed as much in the aftermath of the oil shock of the 1970s and early 1980s, and he told his fellow Muslims that 'the main resource to depend on after God is oil.'"[7]

It appears the "school of hard knocks" is in the middle of teaching the West (and Christians) a lesson. The West perceives its actions, such as sending troops to Iraq, as necessary for our national security by means of removing dictators and promoting democracy in a crucial part of the world. What many in the Middle East perceive as the heavy-handed geopolitical and military actions of the United States and Europe have bred a backlash of resentment and downright hostility to the Western world among many Muslims. The rise in the number of terrorist attacks (and their intensity) proves the despair in large sectors of the Islamic community. For the last hundred years or more, many in Islam have felt the West has trampled their religion and their culture. In retrospect, this is true—indirectly. As the economic and political importance of oil grew after World War II, the West (Christians included) trampled the Persian Gulf region in order to pursue economic and military superiority over Communism. As a result, Islamic culture suffered in two respects: they were unprepared for the tidal wave of Western influence, and their government leaders were ill-prepared for the sudden financial gains, which bred corruption as much as it did religious benefits and social gains.

Oil is truly a double-edged sword. On one side, oil opened the door to a world Islam was not prepared to meet, and as a result, they struggled to resolve their religion with Western capitalism. On the other side, oil provides Islam with an opportunity to write its destiny by using oil to boost Islamic power and influence throughout the world.

Petroleum is undoubtedly the vehicle to resurrect trampled Islamic society, and many, especially the young Muslims attracted to radical forms of Islam, intend to drive the "oil sword" through our Western infidel hearts. In *The Hydrogen Economy*, Rifkin also writes,

> Oil, the energy that helped make the West the unchallenged economic, political, and cultural force in the 20th century, could now become its undoing at the hands of an Islamic world determined to turn the tables and restore its former status as the world's spiritual and cultural arbiter. Of this much we can be sure: Oil and Islam are inseparably linked. The fate of one will, to a great extent, determine the fate of the other in the coming century.[8]

The Osama bin Ladens, Saddam Husseins, and Mahmud Ahmadinejads (Iranian president) of the world are counting on that. It remains to be seen if moderate Islam can wrestle control away from the Islamic radicals who have hijacked their religion, just as Al Qaeda hijacked planes on 9-11.

Many have asked following 9-11, "Why do bad things happen to good people?" One should note that biblically, God sometimes works through evil events and evil people. That does not mean God created evil or inflicts evil on those we wish, but that He may allow evil to happen as a consequence of humanity's poor choices, made of our own free will. However, God can take warfare, political conflict, and opposing nations and use them for His greater good. This could be one of those moments. It is important to remember that all nations are subject to God's power, even if those nations are not founded on Judeo-Christian principles.*

For example, Jerusalem fell to the Babylonians in 586 B.C. because the Israelites had turned from God—not because the Babylonians were any better in the eyes of the Lord. But God used that tragic event for His greater good: to break Israel from their detestable cycle of idol worship, which jeopardized their chance at a relationship with the one true God for eternity.†

---

* "Everyone must submit himself to the governing authorities, for there is no authority except that which God has established. The authorities that exist have been established by God" (Rom. 13:1).

† Habakkuk 1–2 foretells of the coming Babylonian conquest. Jeremiah also deals with the moral decay of God's people at that time, and the impending judgment God would bring by the hand of the coming Babylonian army. Jeremiah also includes tinges of hope about future redemption and the healing of God's people. This gives us all hope and relief that, no matter how bad a situation might get, God can still use it for His greater good.

Today, there is no peace in the Middle East, and two wars in two decades have been waged in Iraq. Where was Babylon located? In present-day Iraq. It's amazing how little has changed in 2,500 years. Are we witnessing the demolition of our petroleum crutch via terrorists and governments sympathetic to radical Islam, and re-focusing our relationship back to God, where it belongs?

Separation of Mosque and State?

Western society has a lot of trouble understanding the Islamic influence on governments and businesses with large Islamic populations. In large part, this is because of how Christians view the world and, as a result, how Western democratic governments are constructed.

The United States, for instance, is a constitutional republic. No matter how much today's secular humanists would like to rewrite history, America's founding fathers were mostly Christian and based the foundations of our country on Christian ideals.[9] Our earliest documents as a nation confirm that fact, especially the Declaration of Independence. Many misinterpret the concept of the "separation of church and state." We hear that phrase so many times that we assume those words are in the Constitution, but they are not. The founding fathers never intended to keep God out of government; obviously we cannot tell God what to do. They intended to keep government from imposing a national religion on the people.

To understand Christian acceptance of secular governments in Western society, you need to understand the Christian mindset. For Christians, our existence here on earth encompasses a very short time compared to eternity. Our Christian experience instills in us a spiritual need to prepare for the afterlife and, as a result, we tend to tolerate things in this imperfect world—perhaps more than we should. Most Christians would agree that we sow the seeds of Christianity, but God determines when and which seeds He will water and grow.* Although some Christians might be overly aggressive in proclaiming the good

---

* Parable of the Growing Seed: He also said, "This is what the kingdom of God is like. A man scatters seed on the ground. Night and day, whether he sleeps or gets up, the seed sprouts and grows, though he does not know how. All by itself the soil produces grain—first the stalk, then the head, then the full kernel in the head. As soon as the grain is ripe, he puts the sickle to it, because the harvest has come" (Mark 4:26–29).

news of salvation in Christ, most understand that we cannot impose our beliefs on those unwilling to accept Jesus. Likewise, in government affairs, most Christians do not expect or want a government that imposes religious doctrine, even if it happens to be their own. Most Christians feel quite comfortable with a democratically elected government, as long as they can practice their religion freely, without having anti-Christian ideals imposed on them.

The same separation of government from religion remains much less likely in Islamic culture. To our surprise, many people of Islamic faith don't want a "separation of mosque and state." In many respects, Islam's inherent practice of striving to imbed their religion in their everyday lives is admirable; something Christians espouse too, but in practice our business culture and laws restrict. But in keeping practice of Islam foremost in their government, the acceptance of other religions becomes difficult.

Although many Muslims claim that Islam tolerates other faiths, the proof is hard to see today in Islamic governments from Morocco to Bangladesh.[10] For instance, Saudi Arabia, home of the two most holy shrines in the Islamic world (Mecca and Medina), has a population that is 100% Islamic.[11] How can that be? Saudi law is based entirely on the strict Salafi (Wahhabi) sect interpretations of Islamic teaching. This prohibits the practice of any religion other than Islam. Throughout the Middle East, the story is much the same; the law or the cultural pressure is so overwhelming that being anything other than Islamic is implausible, if not criminal.

Just ask Abdul Rahman. In the spring of 2006, Rahman was turned in by his family to authorities and convicted of a capital offense in his native Afghanistan. His crime? Converting from Islam to Christianity. He faced the death penalty if he did not denounce Christianity. International pressure spared Rahman's life. A legal loophole in Shari'a law—declaring him mentally incompetent—allowed him to go free.[12] And many would agree that you'd have to be crazy to confess being Christian in a fundamentalist Islamic country! Before we rush to judgment, Christians have done the same in the past. When the Spanish peninsula was re-conquered by Christians in the Middle Ages, many Muslims were violently put to death if they did not convert to

Christianity. Either way, forced adherence or forced conversions are not right.

The fundamentalist Islamic faith teaches complete integration of religion in all facets of life, including government affairs. That core belief makes it difficult for many devout Muslims to understand why the U.S. values democracy and freedom so much, when it potentially erodes our faith. They point to the excesses of Western society—such as drugs, pornography, crime, greed, and homosexuality—as evidence of the moral decay of Christian society. They ask themselves, "Is this what democracy and freedom bring? If so, then we want no part of it." It makes sense that they have no problems picturing us as "The Great Satan."

In part, this depiction of Western excesses is unfair, since the freedom and openness which Western society espouses also exposes our faults, placing them in plain view. Many of these human faults are pervasive in other societies and religions, just not as exposed for all to see. Nevertheless, Christian society has to admit that we need to do more to bring Christianity into our everyday lives. We need to be more than "Sunday Christians" who deposit our morals at the church exit door until the next week.

This is not a new problem. I watched a television program showing classic film footage of Rev. Billy Graham speaking on June 9, 1957, in which he noted that we (Americans) are a church-going nation, but we are not relating church-going to our personal, daily lives. We are not going into our homes, shops, offices, and businesses and putting Christ into effect. We need to apply Christ in our daily lives and social intercourses.[13]

Although it has not happened to date, that call continues today, and Christians should not be discouraged to continue this important spiritual endeavor. However, this call to put Christ back into our public life, our business life, even our government, does not mean that Christians espouse a theocratic national religion based solely on Christianity. It means we need to recognize that the fabric of a nation and its government is its people.

Abraham Lincoln expressed it beautifully in the Gettysburg Address, ". . . government of the people, by the people, and for the people."

Government exists for the purpose of protecting its citizens and providing a framework for getting along with each other. Simply, government is about people, and most people (in the United States and elsewhere) consider themselves religious (or at least spiritual). More than 81% consider themselves religious in the United States; 77% of them are Christian.[14] Therefore, the personal faith that drives many individuals involved in government service deserves inclusion, not exclusion, as one of the many facets to consider when running a nation. Why exclude something so important and vital to our identity as a nation? Have you ever made a recipe and forgotten to add a key ingredient? The result is something usually worse, not better.

In fact, concerning religion, if both the West and the East would move toward the center, both our societies would be better off and understand each other better. The West should be more inclusive of religion, recognizing that it is part of what defines us. Governments in Islamic countries should live up to what Islam claims by accepting other religions more freely.

If water were the religion in the pool of our government, the West would not have enough to swim, and the Islamic world would drown in the deep end. We need a pool where we can swim and stand, buoyed up by our religion, but not drowning in it.

Likewise, if we filled the pool with oil instead of water, the West would not have enough to swim to economic fortune, and the Middle East would drown in excess.

Explosive Problems

The oil boom of the past half century initiated a side effect in the Persian Gulf region, especially Saudi Arabia—explosive population growth, largely Islamic. The median average age of a Saudi citizen clocks in at an alarmingly low 21.5 years,[15] compared to 37 in the United States.[16] Almost 40% of Saudi Arabia's population is 14 years old or younger.[17] Any time a society's population booms with numbers like that, it is almost impossible to educate and create good jobs for everyone. A substantial number fall behind, especially when a good part of the population is financially poor to start.

The arable land Saudi Arabia possesses to feed its booming population is less than 2%.[18] The lack of adequate, clean water supplies presents perpetual problems. There are no industries to speak of outside the petroleum and natural gas industries, which are not particularly labor intensive. Most of the other existing industry derives from natural resources. Government social programs, which support a tremendous portion of the population, strain under the burden of the population explosion. Disenchanted youths in Saudi Arabia resent the fact that their fate rests on the price of crude oil, which supports Western civilization. High oil prices up until 2008 have filled the Saudi coffers again, easing some of the economic tensions, but the war in Iraq, Arab-Israeli tensions, and trouble in other parts of the Islamic world have other kept social tensions high.

It's easy to see why the youth of Saudi Arabia are prime targets for Islamic extremists. The 9-11 attack on the twin towers of the World Trade Center included nineteen hijackers, fifteen of them Saudis. Recently, terrorist activity internal to Saudi Arabia rose sharply, including, for the first time, an attempted attack on a major oil pipeline in the spring of 2006. The Saudi government has only recently admitted the existence of these homegrown terrorists in their midst.

Beyond the terrorist issue, the sheer logistical problems associated with the population explosion continue to plague the government, since it provides many of the nation's jobs. Low oil prices in the late 1990s fueled rampant unemployment, causing a swell of grass-root extremist movements. Now that the oil economy has recovered, large numbers are still angry—and well-financed. The Saudis must soon find a solution to break the population's dependence on the government and establish a stable economy less dependent on oil and natural gas.

The Muslim population boom is most pronounced in Saudi Arabia, a relatively wealthy country, but the Islamic population is also the fastest growing around the globe. The decline in the economic health of the Islamic world for those countries without natural resources further increases the unrest in Islamic society.

For instance, the most populous Islamic country, Indonesia, left OPEC at the start of 2009. They are rapidly becoming an oil-importing country. Indonesia suffered from the economic collapse in Southeast

Asia in 1997, along with the rest of the region. The population growth, coupled with the economic decline, brews a recipe for disaster. The "rich" Islamic countries, dealing with wars, geopolitical instability, population booms, and the like internally, have little to spare to bail out their Muslim brothers. The greed of the West becomes an easy and believable target on which to pin Islamic troubles, and therefore too many Imams (Islamic teachers) incite their fellow Muslims to react with violence. One thing is certain; the internal problems of Islam will not be solved by external violence. Neither can the internal problems be appeased by a flood of oil money.

The (Not Too) Funny Papers

Islam claims to be a peaceful religion. In recent history, it has lost much credibility in that regard. If Islam wants the rest of the world to start taking it seriously again, then events like the Islamic rioting which occurred after the cartoon depictions of the prophet Muhammad were published in a Danish newspaper must stop. Are cartoons, even offensive cartoons, worth taking human life and destroying millions of dollars of property?

There appears to be an ever-widening gap between the level of respect Islam shows to other religions (not just Christianity) and how Islam demands to be respected. If Islam is a religion of peace and compassion, then moderate, peaceful Islamic leaders must start taking active measures to counteract the radical Islam movement. So far, it's hard to see evidence that this is occurring.

For instance, some Muslims in the region where Osama bin Laden is hiding must know his location. Why have they not turned him and other al Qaeda operatives over to authorities? Actions and words appear to be out of sync. Similarly, in the case of the cartoon response, the voice of moderate Islam denouncing the violent actions of the Islamic perpetrators was only a whisper. If this ineffectiveness of moderate Islam continues, the outside world will see only an outward expression of internal hopelessness within Islam. Radical Islam is persecuting its own proclaimed religion, not just Christianity and other world religions.

Conflicts within Islam

Besides the internal battle for control of Islam between radical and moderate factions, Islam has centuries-old divisions within—namely the Sunni and Shia branches. Sunni Muslims are the largest sect, with about 90% of the Islamic population,[19] while the Shia (or Shiite) branch is prevalent in the Iran/Iraq region. The split formed immediately following the Prophet Muhammad's death, with some following Ali, Muhammad's cousin (which became the Shia branch), and the remaining following a chosen leader or "Caliph," which became the Sunni branch. Today, Islam struggles with disturbing violence between the Sunni and Shia branches.

Granted, Christianity has also struggled with this. The bloodshed between Christians during the Reformation, and the Catholic versus Protestant bloodshed in Northern Ireland, shows we have strayed from God's will from time to time. However, Christianity presently remains, for the most part, at peace with itself.

What causes the present internal discord between different sects of Islam? Is it because one side is siding with the Christian West and the other disagrees? No, both sides have strong anti-Western sentiments within large segments of their branch. Is only one side militant and the other not? No, that's not the answer either. Does only one side have oil and the other not? No, although only 10% of Muslims are Shia, the Shiites dominate the populations of Iran and Iraq, who possess huge reserves of oil.

Nevertheless, Christians should care about the rift between Shia and Sunni Muslims, not only because it further destabilizes the Middle Eastern oil reserve picture, but because it also makes it all the more difficult to show the peace found in Christ to them. Christians don't need to choose sides with either sect, but we do need to understand that there are differences.

No Jesus, No Peace; Know Jesus, Know Peace

I saw this painted on the side of a shed in the middle of Kansas and I couldn't help but think of how true this is in the Middle East. That is why it is so important to remove the codependence on oil that Christians and Muslims share. Christianity will never be able to honestly deal with Islam—and Islam will never be able to deal honestly with Christianity—as long as the power of oil dictates our actions towards one another.

You can see it in the way Big Oil business interests and Western governments pander to totalitarian governments with abysmal human rights records and corrupt regimes, both past and present—all in the pursuit of oil. You can also see the Islamic dependence on oil for cash. Some of the poorest peoples on earth live in Islamic regions. All the money from oil has not helped them.

Although the Middle East is half a world away, we share responsibility for the environmental health of the planet and interact with each other much more today than a few short decades ago. And like it or not, we are more intertwined today with our Islamic neighbors in the matters of global economics, politics, and environment largely because of oil. With air travel, the internet, cellular communications, supertankers, and other modern marvels, the global economy is a reality. Ask any three-year-old who has visited Disneyworld, and he or she can even sing to you, "It's a small world after all."

But to our dismay, it's not a peaceful world. Cultural differences and hundreds of years of relative isolation between the Islamic world and the West make mutual understanding difficult to achieve. These differences will not disappear overnight, but by the grace of God they can be overcome. Part of the healing happens when we realize not all Muslims are terrorists, as too many believe. There are well over one billion Muslims in the world. If they were all terrorists, we would be in global war down to every last man, woman, and child.

My personal experiences with Muslims testifies to their kindness and generosity. To vilify them completely goes against everything Christ teaches. However, that does not mean God wants us to accept their beliefs as equal to ours. How then are we to act?

We need to be patient with them, just as God is with us. Patience is the opposite of judgment; we cannot do both simultaneously. Forgiveness and mercy stem from patience (Rom. 2:1–4). We try to take the power of God in our own hands when we judge others. Just because God has allowed terrorism to persecute us through Islam does not mean we have the right to judge and condemn them. We are sinners just as they are. God's judgment will come according to His timetable, not ours.

Beth Moore states it clearly in her Bible study, *Living Beyond Yourself:* "We cannot both judge others and be patient toward them . . . To judge why God won't bring a speedy punishment to those who mock or despise Him is not our responsibility. To all of us, His long-suffering has been our salvation. Eternity is a long time: the effects are irreversible. God desires to give every chance . . ."[20]

Until God's judgment, we should continue to proclaim Jesus as our risen Lord and Savior, defending the truth of God's Word, and allow, through patience, the Holy Spirit to work through us.

CHAPTER 4

THE BIG "E":
ELECTRICITY

My wife, my son, and I sat with our candles, the light from our flashlights growing dim. Hard to believe only three hours had crawled past since the power went out and catapulted us back to living an 1850s lifestyle. My wife, who I tease about having a one-and-a-half degree comfort zone between "too hot" and "too cold," wrapped herself in a blanket like a burrito in front of the fireplace, the one supplemental source of heat in our all-electric home. My son thought it was pretty cool and exciting for the first twenty minutes after the power went out. Now he was quickly growing impatient with the lack of options, even my suggestion that he go to bed early. I, too, was growing tired of the inconvenience, but tried to suppress my dissatisfaction. I had things to do. Sure, the first two hours of the power outage were somewhat relaxing, but now I longed to get back to normal.

On the hunt for fresh batteries, I found my way down the dark hallway and flipped the light-switch on as I entered the room.

"Hey, goofy," I muttered to myself, "don't you know the power is out?"

Then I realized that 95 out of 100 people would probably have flipped the light-switch out of habit, the same as I had done.

At this point, the light came on (in my head at least) as I realized how much we take electrical power for granted. As consumers, we often remain oblivious to the importance of electricity—from how it's generated to how it's delivered to us—until it fails. Only then do we appreciate its effect on our lives and businesses. As with gasoline, we're dependent on electricity, but pay no attention to it until we don't have it, or until prices rise and our pocketbooks start emptying.

Western civilization would cease to be civilized without electricity. Just ask anyone who experienced the August, 2003 northeastern blackout, which left 50 million people without utility power for hours and caused an estimated economic loss of $6 billion.[1]

Think of the refrigerators, computers, traffic lights, heating and air conditioning units, elevators, and other electrical devices we use daily. We expect reliability from electricity, just as we expect a chair not to collapse when we sit down. We realize, at least temporarily when faced with the hopelessness of a blackout, that electricity is important. When it is unavailable, we realize that we have neglected it. At those moments, we discover that we never want to lose it again.

Truth is, about a third of the earth's population does not have electricity. About two-thirds of the earth's population does not know the true Jesus. We unfortunately take our faith for granted as we take electricity for granted. Of course, not taking our Christian faith for granted is more important. However, considering the importance of electricity to us (even if it is underappreciated), we as Christians should seek to practice good stewardship of God's creation by considering carefully how we generate that energy. We have many options. The Christian position does not necessarily say, "God only approves of solar power" or "Jesus only loves hydrogen power." Most people, including Christians, would love to believe "God approves of cheap power." We should seek wise application of resources to generate power cleanly, not just efficiently, always striving to make the right choices at the right time. We must also act accordingly when God-given opportunities present themselves to help the poor improve their lot.*

---

* Electricity is one of the needs of the poor we should supply along with medicine, housing, and schools. It represents a building block that creates possibilities. Think of it this way: if you were so bitterly cold that your teeth chattered, or hot to the point of heat stroke, or hungry to the point of starvation, or if disease ravaged your body, you would beg someone to care for those needs first. It's so hard to reach someone spiritually if you haven't demonstrated your concern for their physical needs first. Electrical power—for pumping water, for medical equipment, for air conditioning, for heating, for light—is extremely helpful.

Let's spend some time learning about electricity, from generation methods to power delivery. Take a look at the simplified diagram below to see how electricity is typically generated and transmitted from beginning to end.

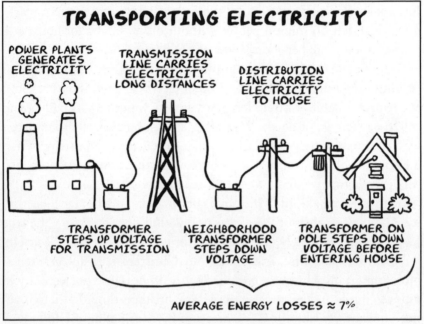

Figure 4.1—Simple Electrical Generation, Transmission, and Distribution.

Let's get a few simplified electrical terms and concepts under our belt:

Volts

The term voltage is a measure of electrical potential energy, or the amount of force available to push electrons through a conductor wire. For example, an orange on a table has potential energy—if it rolls off, it falls to the floor. An orange on the floor has lost its potential energy since there isn't any distance for it to fall any longer. A battery has potential energy to power a cell phone until the battery is discharged. Recharging restores its potential. For a vacuum sweeper, the wall outlet has 120 volts of potential energy to run that sweeper, but it won't do so unless it's plugged in and turned on. Only then is its potential utilized. Voltage can be DC, such as a 12V DC car battery, or voltage can be AC, such as our 120V AC home wall outlet.

Amps

The term Amperes (Amps for short) is a measure of current. Current can be thought of as a stream of electrons. Electrical current is analogous to a stream of water. Water in a stream runs from higher ground to lower ground. Without that difference in ground height, there would be no current, or flow of the water. Likewise, without an electrical potential (voltage), there can be no current, or flow of electrons. The term DC means "direct current," a constant flow of electrons, while AC means "alternating current," which means the electrons change direction repeatedly in their flow down a wire in sinusoidal motion. To understand sinusoidal motion, picture in your mind how a snake moves forward. Its body slithers back and forth in an "S" pattern, similar to how the electrons move forward inside the wire.

Watts

Watts are a measure of electrical power. In simplest terms, watts equal the voltage multiplied by the current (Volts x Amps = Watts). In electrical generation terminology, it is common to state power plant output ratings in Megawatts (MW), which tells how many millions of watts the generating station can produce at any second in time. The flip side is the rated consumption of watts. This represents the number you see when buying light bulbs. A 100-watt incandescent light bulb consumes that much energy per second.

This really isn't the best way to "size" light output, but it is the way the industry developed and now everyone thinks of light bulbs in those terms. That's why when you see a "100 Watt" compact fluorescent light bulb (CFL), it will also have something like, "consumes only 27 Watts of power" on the packaging also. The CFL bulb is designed to give the light output equivalent of the 100W incandescent bulb people are accustomed to buying. Lumens, the measure of light output, would have been a better term to embrace, if we could turn back the clock.

Energy Storage Aspects of AC and DC Power

A critical concept to understand, and one that will only become more important in the future, is the energy storage aspects of AC and DC power systems. Energy storage exists in many forms, (including fossil fuels and uranium), but AC electricity is not one of them.

The type of power put out from electrical power plants is AC. For technical reasons related to physics that I won't bore you with, AC electricity must be used immediately at the time of its generation output from the power plant. AC stands for "alternating current," and when it stops alternating, the electricity disappears immediately, not gradually. An example that illustrates our inability to store AC electricity can be observed when power to your residence goes out. In less than one second, television sets turn off, lights go out, and desktop computers crash. If any electricity were stored in the system (in the power plant, power lines, or your residence) the AC power loss would not result in instantaneous power loss. With DC (direct current) electricity, such as that supplied by batteries, power loss is gradual, as observed by a flashlight growing dim as the batteries get weaker.

The energy storage aspects of AC and DC power systems are unique—analogous to cats and dogs; both are 4-legged house pets, but they are unique. AC power possesses benefits and drawbacks. So does DC power. Unfortunately, in most cases, AC and DC power cannot be used interchangeably. Coexistence of both types of electrical power will continue, with each applied where it is most beneficial.

You may think, "Wait a minute; why not use batteries to store electrical energy and use DC power exclusively?" But have you ever seen a battery provide normal power to a house, office building or town? Probably not, since it is cost-, size-, and weight-prohibitive to store large amounts of power in battery systems, plus the durations are too short before the battery needs recharging from—you guessed it—AC electricity. Battery systems are better suited for small power applications, like backup power and auxiliary power. Batteries store DC power only, and although it can be converted to AC power, it's not practical to do so for large-scale power generation. In fact, the largest battery system in existence provides power equivalent to 27MW of AC power, but can only do so for approximately 15 minutes; enough time to connect other AC power systems to the grid and return to normal operation.[2]

Watts-Up with Electricity Generation?

Electricity generation and transmission originates from somewhere, usually a distant power plant, since most citizens do not want a 600-megawatt generating station in their backyard. Power plants are noisy, unsightly, and too utilitarian to be considered beautiful, even by those who design them.

So why should we care how we generate electricity? Because the air we breathe, the water we drink, and other physical necessities are linked to electrical generation. Great strides have been made in reduction of power plant emissions, but we need to do more. Emissions from fossil fuel power plants affect the air we breathe. These include carbon dioxide (which affects the rate of our climate change), nitrogen oxides (which form smog), and sulfur dioxide (which forms acid rain). Airborne mercury poisoning of lakes and streams from coal-fired power plants contributes to serious health concerns. Water emissions from fossil fuel and nuclear plants affect the surrounding environment. On-site storage of spent (used) nuclear fuel, transportation of nuclear waste, and long-term storage are valid social concerns, even if a record of accomplishment of safety is maintained.

Yet with all these concerns, we cannot do without the electricity that fossil fuel and nuclear power plants provide. The social and economic impacts would be enormous for both rich and poor. Besides, even if we wanted to change all fossil fuel and nuclear plants with something else, the process would take decades. That does not mean we should let the status quo continue unexamined; it means colossal changes to electrical infrastructure do not happen overnight.

As to how we generate electricity, let's look at the source percentages for electrical power in the United States.[3] The sub-group percentages add up to main group total percentage. Some of the technologies in development stage are not tracked individually; therefore, no percentage is shown. There are a number of ways of generating electricity, as shown below, which relates to the left side of the diagram (figure 4.1) included earlier in this chapter:[4]

1. Fossil Fuel Power Plants—71.7%
    a) Coal—48.5%
    b) Natural gas (Methane)—21.6%
    c) Oil- 1.6%
2. Nuclear Sources (Fission)—19.4%
3. Alternative / Renewable Technologies—8.5%
    a) Hydroelectric—6%
    b) Biomass (wood, wood waste, and other biomass)—1.3%
    c) Wind—0.8%
    d) Geothermal—0.4%
    e) Solar (Photovoltaic / Thermal)—0.05%[5]
    f) Ocean (Tidal / Current / Wave / Thermal)
    f) Biological / Chemical
    g) Batteries
    h) Fuel Cells / Hydrogen

There are others, but the main technologies are all included. As you can see, more than 90% of all of the electricity generated in the United States comes from fossil fuel and nuclear resources. One can look at the different electrical generation technologies as legs on a chair. The chair is hard to balance when the legs are uneven.

The reason for the lopsided percentage is simple—fossil fuels and nuclear sources have been the cheapest method of producing electricity to date. Is that economic advantage guaranteed for the future? The proponents of energy diversity would say no. Many experts confirm it looks more and more like fossil fuel's economic dominance is ending because of volatile prices, which are rising faster than other technologies in general, and emerging environmental cost impacts. Nuclear power is on an upswing, but not cheap. Alternative energy technologies often cost more than fossil and nuclear fuels, but the price gap is closing between alternative energy technologies and the others. In some economic analyses, some alternative energy technologies already cost less. Even from a purely economic standpoint, one can make a strong case not to put all our eggs in one basket, leading to the increasingly popular mantra of *energy diversity.*

Dividing Up the Pie

Let's take a closer look at the options available, both traditional and new. What's the best mix of energy technologies, the energy diversity we'd like to see? It varies from region to region, country to country, and economic model to economic model. The various methods of generating electricity have distinct characteristics that we'll explain to give an educational background, without getting too judgmental. Plenty of reasons to diversify our generation of electricity exist:

- Better environmental balance—cleaning up the air and reducing greenhouse gases
- Price stability—not relying on any one energy resource too heavily mitigates economic impacts of fuel disruptions
- Increased sustainability—by increasing renewable technologies and extending the availability of our fossil fuel resources for future generations
- Increased job creation and economic growth—brought on by developing new industries such as coal gasification, wind power, solar, fuel cells, and the like. These grow good-paying technology and manufacturing jobs due to the increased complexity of equipment production and implementation over traditional technologies.

Coal

Coal is the most abundant fossil fuel in North America. That is a plus from economic and national security respects for our electrical power system. Some estimations project the United States has enough domestic coal reserves to last more than 150 years.[6] Foreign governments and terrorist groups hostile to the USA cannot control or affect our economy with coal to the extent they can with oil or natural gas.

Coal-fired power plants, as a mature technology, boast readily available equipment, well-refined codes and standards, and an experienced work force which can design, maintain, and operate the plants. New technologies like Integrated Gasification Combined Cycle (IGCC) are improving the environmental picture for coal, but are still expensive and not yet in widespread use. Coal-fired power plants make good *base load* (continually operating) plants for two reasons. One, since coal-plants require hours to purge and restart, it's more cost effective to keep them running, and it's also easier on the equipment. Two, economies of scale are better for large plants. Large power plants create the desired power much more efficiently, because smaller plants typically waste a higher percentage of energy to heat loss and other process losses.

Opposite to base load plants, *peaking* plants only operate when the load demand for more electricity requires it, like during the afternoon hours of hot summer days. Peaking plants are easier to start and stop, something definitely not true of coal plants.

Put another way, *base load* is like the amount of food that would be required to feed a family. *Peaking load* would be the added food required to feed the hungry neighbors who joined in for dinner. While we are at it, *spinning reserve,* which is the extra electricity generated to accommodate unexpected power needs, is equivalent to the amount of food a good host would provide his family and guests to have a second helping. For both power supply and dinner, the trick is to have enough to satisfy, without having a bunch of leftovers.

If handled improperly, raw coal and its post-combustion byproducts can make coal a major source of pollution. Proper handling of water runoff from coal storage and coal washing is crucial so that the dirty water runoff does not end up contaminating waterways and groundwater. After combustion, ash byproducts must be disposed. Some industries utilize some of the ash byproduct. For instance, concrete often includes ash as one of its components. Nevertheless, the quantity of ash often exceeds external uses, and ends up as waste.

What's worse for the air are the flue gas emissions that rise from the chimneys of a power plant. Technology and regulations in developed countries have greatly reduced the percentage of emissions per pound of coal burned. These emissions include nitrous oxides, acid-rain-causing components like sulfur oxides, and other particulate matter. In 2005, the United States became the first country to enact emission restrictions on mercury in flue gas. Mercury is a poisonous metal that settles in lakes, streams, and oceans from the flue gas and contaminates fish if the concentrations get too high. Since the mercury is not processed by the body of animals, it accumulates first in the bodies of fish and then later in the bodies of humans who eat the fish. Reducing these emissions is a step in the right direction, until cleaner base load sources of power prevail.

Another problematic air emission for coal-fired power is carbon dioxide. Carbon dioxide is a greenhouse gas which contributes to climate change. Presently, carbon dioxide is not under mandatory restrictions in the United States by specific law, although the Supreme Court has ruled that the EPA has the power to regulate carbon dioxide as part of the Clean Air Act. Specific legislation to encourage carbon restriction is only a matter of time, as is "price" assigned to the carbon.

The increased use of coal, especially in developing countries such as China and India, offsets technology improvements and regulatory restrictions; therefore, overall emissions of all combustion byproducts continue to climb, although at a lesser rate.

Coal already accounts for a large portion of our electricity generation, limiting our energy diversity and making clean air standards harder to maintain. Relying on any one particular energy generation technology is probably a bad idea because of the disaster it could cause if problems occurred in that particular industry. For example, the prolonged shutdown of the railway system by labor strike or multiple terrorist strikes at key locations could adversely affect coal plants. Coal-fired power plants sometimes only maintain a couple days supply of coal and rely on timely coal deliveries. Since a coal-fired power plant often operates twenty-four hours a day, seven days a week, the off-peak hours with minimal electricity demand often undesirably adds to the spinning reserve. Although this type of operation is inefficient during low loads, it is better overall than shutting the coal-fired power plant down for short periods like a day.

Concerns for safety also develop because of the highly explosive coal dust created from processing and handling the coal. Special equipment helps to mitigate those risks. Even with modern technology, however, mining the coal to use in the power plants remains one of the most dangerous professions in the world, reaffirmed by the West Virginia Sago Mine disaster in early 2006, where eleven coal miners died after being trapped by an underground methane explosion. The dangers are even greater overseas, where thousands die every year in mine-related accidents.

Few people want a coal plant in their backyard; therefore, we typically build new coal plants in remote areas. This increases electrical transmission power losses because of the increased distance. Coal power plants require large quantities of fuel, so they must be closely located to railroad lines, large freshwater waterways, ocean ports, or right outside the coal mine. These "siting" problems are becoming more and more prevalent because of government regulations, environmental concerns, and quite often, local public opposition.

Natural Gas & Fuel Oil

Until recently, when natural gas prices spiked after damage caused by hurricanes Katrina and Rita, natural gas-fired combustion turbines were among the cheapest electrical power producers. Now many realize that natural gas pricing is as volatile as petroleum pricing. Likewise, it unfortunately shares many of the same foreign supplier problems as its fossil fuel cousin, crude oil.

In the 1990s, natural gas was king, and almost all new fossil-fueled power plants built were combustion-turbine based. A combustion turbine is the big sister to the jet engine on a large airplane like a Boeing 747, except combustion turbines use natural gas (methane) instead of jet fuel and are usually at least triple the physical size and power of a jet engine. Instead of attachment to an airplane wing, the combustion turbine is mechanically coupled to an electric generator to produce electricity.

On the environmental plus side, it's the cleanest power provider of the fossil fuels. Because natural gas is methane (CH_4), it contains the least amount of carbon atoms in the fuel. Therefore, when combusted, it produces less carbon dioxide, carbon monoxide, and other air emissions. Coal and petroleum contain more complex hydrocarbons with more carbon and sulfur molecules, so they cause more troublesome emissions when combusted.

Natural gas has other industrial uses and residential heating uses that outweigh its use in electrical generation. Experts estimate that natural gas supplies will last longer than oil, but it's not as plentiful as coal. For these reasons, over-build of combustion turbine power generators should be monitored closely.

Nuclear

Quite a few people are surprised by the number of nuclear power plants operating in the United States and across the world. France receives about 77% of its electricity from nuclear power plants,[7] compared to 19.7% for the United States. A little over a hundred nuclear reactors in the U.S. supply that power, but because public perceptions changed after accidents at 3-Mile Island on the east coast and at Chernobyl in the former U.S.S.R., new nuclear plant plans virtually stopped in the United States for over thirty years.

Nuclear is trying to make a comeback, supported by recent regulatory policy changes, the Energy Policy Act of 2005, and a changing political and social climate. Since nuclear plants do not produce dangerous air emissions, even some environmental groups support nuclear power as the lesser of the two evils when compared to air emissions from fossil fuel plants.

Considering the amount of nuclear electrical power produced since the 1950s, the industry safety record is actually quite good. The problem lies with the potential for devastating disaster. Increased terrorist activity and America's vulnerabilities exposed by the 9-11 attacks make many people nervous about nuclear power sources.

Nevertheless, the promise of cleaner air, in conjunction with the reduction of complexity and greater inherent safety of new nuclear plant designs, is convincing many that nuclear needs to remain part of the electrical supply equation. However, spent nuclear fuel still possesses thousands of years worth of radiation. Whether new nuclear plants are built or not, long-term nuclear waste storage issues must be addressed. For instance, we've spent millions of dollars studying and designing a geologically stable and secure long-term repository for all the spent nuclear fuel from U.S. power plants. Yet the storage site proposed at Yucca Mountain in Nevada remains in political limbo, as it has for the last twenty years. Although it came close to political reality in the year 2005, another round of political debate kicked off in 2006 and the project stalled once more.

Because of the chronic political wrangling resulting in Yucca Mountain delays, what was supposed to be temporary storage at nuclear power plant sites has become long-term storage by default. Storage at a nuclear power plant site may solve the long distance transportation problems that many states object to on the grounds that nuclear waste from another state could possibly endanger residents of their state. It is, however, only a band-aid fix to the problem. The used nuclear fuel (called spent fuel in nuclear jargon) requires long-term storage capable of containing it for the thousands of years necessary for some types of nuclear radiation to decay to safe levels. Alternatively, it is possible to reprocess some of the spent nuclear fuel at off-site facilities, making it usable again as nuclear fuel. This is promising since we have few other solutions for spent fuel—and new uranium supplies are dwindling too, like fossil fuel resources.

Alternative Energy

Alternative energy can be defined as sources of energy that are neither fossil fuel nor nuclear. Renewable energy is a sub-group of alternative energy sometimes referred to as green energy. Renewable energy is derived from sustainable, inexhaustible sources, such as wind and solar power. When you consider that only 8.5% of our present electricity supply comes from alternative energy, one must conclude that this is an untapped but crucial opportunity.

Hydroelectric

Hydroelectric is the largest developed energy source within the alternative/renewable energy group. Engineering marvels like the Hoover Dam in Nevada/Arizona are examples of large-scale hydroelectricity generation projects, but other small-scale hydroelectric projects in small rivers and streams are becoming increasingly common.

Interestingly, hydroelectricity, or *hydro* for short, has lost some green energy favor among environmental groups in recent years due to claims of aquatic mortality (fish kill) and other environmental impacts. Although presently there is not much large-scale new development in the hydro sector, it still deserves a place at the renewable energy table as the grandfather of renewable energy.

In 1940, approximately 40% of electricity was generated by hydropower.[8] Canada still generates 60% of their electricity from hydropower.[9] Since most of the best large-scale hydro sites have already been developed in the United States, don't anticipate hydro to save the country from high fuel costs with an expansion boom. Small hydro projects may re-emerge if recommendations from the Energy Policy Act of 2005 are enacted.[10] The greatest impact could be incorporating electric generators to the existing 77,000-plus dams which do not have electrical generating provisions.[11] This is because opposition to new dams is much greater than retrofitting existing dams, or upgrading existing hydroelectric generators with newer and more environmentally-friendly equipment.

WIND

Wind power is the fastest growing energy resource in the renewable energy sector. Worldwide, it has grown more than 30% per year since 2000. The rate of growth is impressive, but because of the vast size of the electrical power industry, it still has a long way to go compared to established producers such as coal-fired plants. In the U.S., wind energy accounts for less than 1% of total electricity produced.[12] Technology improvements and increased turbine size (2 or 3 MW per turbine) have made wind power more economically feasible. This is especially true for large-scale *wind farms*, which are groupings of multiple wind turbines. Wind farms can be located on land or offshore, providing some flexibility in location not possible with other technologies.

Wind power does not pollute the air. And since there are no fuel requirements, there is no cost associated with fuel transportation, fuel burning, or disposing of fuel waste. Wind is a free gift from God caused by the sun heating up the earth's atmosphere unevenly, which causes air to flow. Even before Christ's time, windmills provided power to pump water in China.[13]

Nevertheless, challenges still exist for wind power. For one, the wind doesn't always blow at the time the power is needed. Sometimes the wind blows too hard and the wind turbine must turn off to protect itself. These issues of power output intermittency result in a very significant problem for wind power (and similarly for solar power). A term associated with this intermittent power availability problem is *capacity factor.* Improvements made to the wind turbine designs, along with a good windy site, can result in capacity factors greater than 40%.* But consider this carefully—if your television only worked on average 40% of the time, and you could not control when those working hours occurred, you would be outraged.

Vast improvements have occurred in the science of wind prediction, but this is a problem only solved by incorporating energy storage. It would be ideal to store the energy output from the wind turbine directly as AC electricity, but that is not possible.

To help visualize why AC electricity cannot be stored, imagine the concept of trying to put wind in a box; it ceases to be wind and becomes only air. Likewise, AC electricity cannot be "boxed up" and stored since it's not an off-the-shelf commodity like spaghetti or bananas.

Intermittent wind power, compounded with the fact that AC electricity cannot be stored, makes alternate forms of energy storage very important for wind power. However, while incorporating energy storage may increase the power availability, it also increases the investment cost and system complexity.

People often ask why utilities resist adding wind power onto the power grid. Simply put, reliability is a difficult issue. Electric power systems demand a very steady output for reliable operation. Electrically speaking, disruptions of just a couple seconds are potentially lethal to grid stability. Electric generators are designed and protected by relays that sense and open breakers to protect the generator within fractions of a second. Sudden changes of large numbers of wind turbines turning on or off in a small geographic area can disrupt the system, or at least make steady output a challenge. It is harder to operate and protect the electrical supply system, which could lead to more frequent power losses, even widespread blackouts if disruptions are large enough, compared to other generators in the region. Wind turbines, by nature of their intermittency, are more likely to cause disruption to the grid, affecting system reliability if the other grid generation sources are not robust.

---

* The capacity factor is given as the actual energy output divided by the rated energy output over a period of time, usually a year. A capacity factor of 0.3 (or expressed as 30%) is considered good for wind power. This would mean the site would expect to see conditions good enough to power the wind turbines on average to 30% of the rated energy capacity.

Another question people ask about wind power is the number of bats and birds killed by turbine blades. Turbine blades are especially dangerous for raptor-type birds, such as hawks. However, solutions exist, such as proper location choice or bird sensor shutdown of the turbines when flocks of migrating birds come through. Statistics have shown that house cats kill a lot more birds per year than wind turbines, but the Washington D.C. feline lobby must be keeping that story quiet. Studies have shown that mortality from birds flying into windows and power lines far outweigh deaths from wind turbines.[14] As far as bats go, the wind turbine blades are much harder to avoid for bats and therefore, more dangerous. Many people would say that eradicating bats is a good idea, but unless you are fond of mosquitoes, it would be a good idea to keep the bats around, since they feast on mosquitoes.

Due to the enormous size of individual wind turbines, wind farms also take a large amount of surface area per megawatt generated when compared to a fossil fuel or nuclear plant. However, the land can still be used for farming or ranching even with the wind turbines in place. To increase the economies of scale, larger units are being designed with a goal of 5 MW per turbine for offshore use. However, this requires even bigger turbine blades.

Finally, some critics find the wind farms aesthetically unpleasing and complain of noise from the turbine blades. Wind turbines built in recent years are much quieter, so this problem is not as troublesome as it once was, especially in rural areas where the wind turbines are often located. If proper location suggestions are followed, such as keeping the wind turbines more than one-and-a-half miles away from residential areas, then nuisances are minimized. Proper choice of wind farm site location can make a big difference, but finding a balance between good wind resource availability, away from population centers but near transmission lines, is difficult. The site must be consistently windy enough to warrant a wind turbine, located relatively close to where the power is needed, and situated in a spot that doesn't destroy the view of some of God's more unique and awesome scenery and wildlife. As long as wind turbines don't become as numerous as Starbucks locations, they bring more benefit to society than detriment.

BIOMASS

What is biomass? In simple terms, biomass is organic material from biological origins. So how does that differ from fossil fuels such as oil, natural gas, and coal? Aren't they of biological origin too? Yes, but fossil fuels derive from decayed organic materials in the Earth's crust that have been chemically altered over time by geological heat and pressure. Biomass, on the other hand, has not gone through the geological decay transformations. Biomass includes organic materials such as residual plant crops, animal wastes, leftover wood waste from sawmills, and biogases naturally released from landfills and sewage treatment facilities. Granted, this list includes some pretty nasty stuff, but it's material that would go to waste anyway, so why not put it to good use?

Biomass can be burned by itself or "co-fired," which means burned in mixture with fossil fuels to reduce the amount of fossil fuel used. The heat from this combustion boils water in an effort to create steam, which then drives a turbine generator to make electricity. Co-firing biomass is one of the most cost-effective and easy ways to improve the ecological footprint of a power plant.

This heat-steam-turbine-generator cycle is typical for coal-fired power plants, nuclear power, and other forms of traditional power generation as well. Well over half of the electricity generated in the US and worldwide uses this steam cycle approach.

But there is another way to generate electricity from biomass and the other fossil fuel processes that involve combustion. This other technology is "gasification," which is the process of making a gas from the original coal or biomass materials. This gas could then be utilized in fuel cells, which produce electricity without combustion and therefore with much fewer gas emissions. We'll talk about fuel cells in much greater length starting in Chapter 6.

SOLAR & GEOTHERMAL

Out of the remaining technologies, solar and geothermal are the largest commercially, but still add up to less than 1% of the national total.

Solar power suffers intermittent power availability problems, similar to wind power. It is better in the regard that power demand is greatest during the day, which is when solar availability is greatest. However, cloudy days can throw a monkey wrench in the capacity factor for solar power.

Other problems include the amount of land required, the high cost of materials for solar technology, less efficient electrical conversion, and site location restrictions to suitably sunny areas.

However, with the emergence of a new generation of solar technologies, such as CIGS (copper indium gallium diselenide), we see improved efficiencies and lower prices. Federal tax incentives extended until 2015 and state programs, such as California's Million Solar Roof, provide hope for a bright future.

Much of the present manufacturing technology in solar photovoltaic power (PV) and solar-thermal (focused sunlight reflected off of mirrors to heat a fluid to create steam) processes is foreign, but the U.S. has a chance to leap to the next generation of solar technology manufacturing and still take part in strong solar power growth.

Geothermal is also commercially viable in certain areas. Utilizing steam from the earth's crust to drive steam turbines generates geothermal power. This is considered renewable energy because the used steam from the turbine cycle condenses into water and gets pumped back down into the ground to be naturally reheated to steam again. The challenges of this technology result from the unfortunate location of geothermal sites. Often the sites are not located geographically where electricity needs mandate, or are insufficient to provide the desired capacity. Difficulties exist in determining where to drill steam holes, and chances are that some holes might be *dry holes* with no steam. Also, poor quality of steam can greatly increase the costs of the process cycle and maintenance.

Fuel Cells

Of all the alternative energy technologies in the development or early commercialization stage, fuel cells deserve the most attention because of their exciting potential and wide range of applications. Fuel cells (discussed in more detail in Chapter 6) will be the focus because many consider them the optimal long-term solution. I strongly believe hydrogen and fuel cells are a logical destination. How we get there and how long it will take are the unknowns.

Transmission, Distribution, & Distributed Generation

The electrical transmission and distribution system is often referred to as the *electrical grid*. In reality, *grid* would imply an equally spaced distribution of power lines. This is not the case. The power generation, transmission, and distribution system looks more like an irregular family of spider webs. Transmission lines stretch out from the center of the web, the generation station. Smaller distribution lines crisscross the web, bringing power to the people but not tying directly into the generation station.

Look at Figure 4.1 for a centralized power generation system. Notice the power plants, where generation of electricity occurs, are predominately large plants (greater than 50 Megawatts, often more in the 200MW to 900MW range). The transmission (high voltage) power lines leave the plant and deliver the power, often hundreds of miles, to transformers that convert the high voltage to a lower voltage. The lower voltage is then distributed over smaller power lines, shorter distances to homes and businesses.

Power generation and delivery systems can be set up in a *centralized* or a *decentralized* fashion. The United States and most of the world presently use centralized architecture for the bulk of electric supply. The decentralized option, which I will address soon, is often referred to as *distributed generation*.

The centralized power plant system performs well because it takes advantage of economies of scale in the generation of electricity. However, it loses some of that economic advantage because the greater distance the electricity must travel in the transmission lines reduces power output. The greater distance also increases the chance of power failure due to natural or terrorist events. The greater the distance, the greater the power losses accumulated.

On average, about one-third of the energy content from coal fuel at a generation station finds its way to powering your toaster. A dismal performance, but how often do people stop and think about the lack of efficiency? Overall, consumers of electricity have long since accepted, or more likely, are oblivious to the inefficiencies inherent in central generation. I bet realtors have never heard a request by homebuyers to show them houses near electrical generation plants so that power line losses are minimized. People want to be near schools, grocery stores and churches—not power plants.

Since electricity users understandably do not want to center their lives near larger power plants, the other choice is discrete smaller power plants that are geographically located where needed. This is called *distributed generation.*

Distributed generation is not a new concept. Thomas Edison originally chose small local plants as the design of choice and plan for the future. Somewhere in the early years of electrical power system design, centralized power plants became the norm because of economies of scale.

Although this scheme has served us well in the last century, a move back to more distributed generation may be the answer for several reasons:

- Diversity in supplying a mix of central generation and distributed generation brings energy security
- A greater number of smaller plants makes it more difficult for large-scale blackouts to occur by human error, terrorist plot, or natural disaster
- Reduced power line losses
- Reduced need for unsightly transmission lines that many complain about for aesthetic reasons
- Distributed generation can help isolate or reduce localized power quality problems if applied properly using "smart grid" metering and relaying.

Distributed generation holds the answers to electrical system dilemmas, and it could help with the power quality problems mentioned in Chapter 1 because trouble areas with low-quality power can be isolated from the rest of the grid. The isolated trouble area can then be designed with a system that fulfills the corrective need, thereby improving the whole electrical system.

Computers, microwave ovens, and other modern electronic appliances add irregular disturbances to the grid. The existing system wiring wasn't designed to handle these types of electrical loads. Updating the electrical distribution system, metering and protective relaying, and adding distributed generation would alleviate many of the potential problems that could affect reliability.

Blackouts, Terrorism, and the Aging Grid

August 14th, 2003; August 11th, 1996; July 13th, 1977; and November 9th, 1965: What do these dates have in common? They are dates of the largest blackouts the United States has endured in the last forty years. The North American Electric Reliability Council evolved from the blackout of 1965, which paralyzed the lives of over 30 million people in the northeastern United States and Canada. They promised at that time that they would never allow another large blackout to happen—a promise impossible to keep, although the intentions were good.

These blackouts occurred from natural events or human error. But some experts hypothesize that terrorists could attack our relatively unprotected energy grid at critical locations or hack into the computer operation systems and cause similar chaos. The human and economic toll would be tremendous. In addition, the terrorists would likely get the added benefit of watching us destroy ourselves as some of the worst aspects of human nature expose themselves in full force, such as looting, rape, arson, and murder. (Granted, polar opposite actions such as camaraderie, compassion, servanthood, and unselfishness would also occur, but those stories wouldn't grab the headlines. Think of the aftermath of Hurricane Katrina.).

Speculation that terrorists could instigate a blackout in coordination with a 9-11 magnitude attack would be a devastating combination. In Iraq, the electrical, water, and petroleum infrastructures experience continual attacks from terrorists. We know the terrorists recognize the value and demoralizing impact taking out these infrastructures can have on the public.

The aging grid also needs equipment upgrades to keep up with time or growth in electricity demand, resulting in a precarious situation where disaster waits to strike. Engineers and power delivery companies have done a commendable job of keeping electrical reliability very high. We will not be able to blame them if a tragic event happens, because we haven't done as much as we should to address these known problems. Sometimes you have to pay higher rates to "beef up" and secure the system, but we haven't been willing to pay for it with our utility bills or taxes.

What's the answer?

With all the choices available to generate, transmit, and distribute electricity, which ones (or combinations) provide the best stewardship of God's creation? That's a difficult question. We need some general guidelines, well-rooted in Christian doctrine, that apply to all the different energy scenarios:

- Do no harm to what God has given us to take care of on this earth (1 Cor. 10:23–25, 31, 33)
- Use all that God has created and given us in a manner that honors God
- Care for the welfare of others.

The true bottom line is not the economic bottom line. Other things need to be considered. We need to remember the universe belongs to God. "The earth is the Lord's and everything in it, the world and all who live in it" (Ps. 24:1).

CHAPTER 5

GLOBAL WARMING:
FACT OR FICTION?

I t's 5:45 A.M., fully light outside, and I'm uncharacteristically wide-awake. My wife, Bonnie, and I are enjoying a brilliant, sunny June morning on the deck of a cruise ship off the coast of Alaska, celebrating our 10th wedding anniversary. With a loud *crack*, another massive chunk of ice fractures off the side of the glacier and crashes into the bay. The large crowd on deck "ooohs" and "aaahs" in unison as the swell gently rocks the ocean liner. I cannot believe how close the captain sidled up to this incredible sight. As if performing on cue, another *crack* reverberates, and the scene repeats itself. The only other sounds are the click of cameras, the low rumble of the ship's engines, and the splash of waves lapping against the ship in the otherwise silent morning.

The memories etched in my mind that day were some of the most awesome displays of God's majesty I have ever seen, but also some of the most disturbing. Glaciers have advanced and retreated long before man started burning fossil fuels in mass quantities. However, seeing the calving of icebergs firsthand made me realize global warming could be as disastrous as the Intergovernmental Panel on Climate Change (IPCC) claims. Many leading climatologists believe global warming will greatly increase the likelihood of future disasters, such as sea levels rising and increasing intensity of droughts and hurricanes. If their predictions prove true, we will deeply regret the path we travel today.

With the world population at 6.7 billion, there are 4.7 billion more people alive than in 1922.[1] Many of them are poor and ill equipped to handle drastic climatic change.

On the other hand, a case can be made that denying the cheap and dirty energy sources to the poor, the same ones that built our economy and wealth as a nation, is morally wrong if it delays their climb out of poverty. It seems a cruel hoax, but the cheapest sources of energy are often the most detrimental to the environment. Relegating energy to higher-cost alternatives in developed countries is financially burdensome, but in developing countries, the strain can be financially and physically devastating. Do the immediate needs for basic necessities, such as fuel used to produce food, override the long-term effects of pollution on health and the destruction of the environment for future generations? Balancing immediate needs versus long-term effects is difficult, and not conducive to widespread agreement.

Weather or Climate: Climate Change or Global Warming?

Before we dive into saving the planet, let's make sure we understand some terms. To quote an old weather forecaster's saying, "Climate is what you expect; weather is what you get."

Climate is the long-term average of weather for a region. Weather is the daily, at-the-moment condition. Too often, people claim evidence of global warming because it was hot for a week, or because it hasn't rained in two weeks, or because a hurricane hit New Orleans. Likewise, people also claim a late season snowfall, or a cool week in the middle of summer, or areas of growing Antarctic ice caps as evidence against global warming. Weather exudes dramatic tendencies, but it is a postcard snapshot in time. Climate is more like a three-hour movie.

Other terms bandied about include *climate change* and *global warming*. What's the difference? *Climate change* more accurately describes what is happening, since it involves both warming and cooling changes, as well as other variations, such as precipitation. Literally, *global warming* assumes that temperature only increases, which is not true, since some (but not many) climates have gotten cooler. It's important to remember the big picture: on average the earth is getting warmer, whether by man-made or natural causes. In fact, the earth has warmed about 1.1 degree Fahrenheit or 0.6 degrees Celsius from 1900 until 2005.[2] That may seem insignificant, but it is a global average, masking the more dramatic impacts on critical climate regions.

For instance, this global temperature rise is more pronounced at the north and south polar regions by about double the rate of the global temperature,[3] causing a more rapid thaw of the polar ice caps. The buoyant sea ice and the land ice caps that cover Greenland and Antarctica are both rapidly changing. If too much melting occurs, rising sea levels could potentially flood low-lying lands from Bangladesh to Florida.

The large increase in fresh water from the melting ice also affects the salinity of the ocean water and could drastically change ocean currents. So why should we care? The ocean current "conveyor belt" redistributes huge amounts of energy, which dynamically affects climates of many continents, including North America and Europe. Messing with changes to the "ocean conveyor" could result in potentially drastic side effects due to the potential size and rapid climate effects of those changes. Al Gore states this case in his book, *An Inconvenient Truth:*

> The redistribution of heat from the equator to the poles drives the wind and ocean currents—like the Gulf Stream and the jet stream. These currents have followed much the same pattern since the end of the last ice age 10,000 years ago, since before the first human cities were built. Disrupting them would have incalculable consequences for all of civilization. And yet, the climate crisis is gaining the potential to do just that.[4]

However, the years 2007 and 2008 have cooled the planet's average temperature down some,[5] feeding skeptics' claims that carbon dioxide and temperature increase are not as closely linked as thought. Although the year 2008 was still the eighth or ninth warmest on record,[6] the surface temperatures declined from the recent record year of 2005, despite the fact that atmospheric carbon dioxide continued to climb past 386 parts per million.[7] The scientific fact of global warming faces dispute again, at least in some minds.

According to NASA observations, the sun is going through a weak solar cycle, which could cause the earth's temperature to drop even further. Nevertheless, the amount and rate of melting ice at the northern hemisphere polar region makes the warming trend obvious. One of the contentious questions, however, is why. Many point to a phenomenon known as *the greenhouse effect*, which warrants a closer look.

The Greenhouse Effect

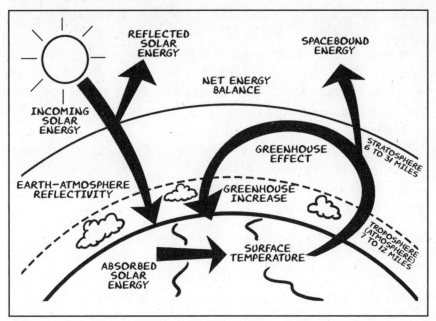

Figure 5.1—Greenhouse Effect, Solar Radiation, and Atmosphere Diagram.

A greenhouse works by allowing sunlight through a glass roof to heat up the interior air. Since this heated interior air is trapped from escaping through the glass roof, the temperature inside the greenhouse is considerably warmer than outside.

The earth's atmosphere acts much the same way as the glass roof. The sun's rays initially penetrate the atmosphere at a short wavelength, but reflect at a longer wavelength from the earth back to outer space. Some of the reflected light and heat escapes back to outer space, but greenhouse gases—primarily water vapor, carbon dioxide, and methane—trap some of the long wavelength reflections. We need a certain level of greenhouse gases; otherwise, the planet would be, on average, thirty to sixty degrees colder, making much of the land a frozen sheet of ice.[8]

"Uncle Al"

The earth's albedo—the fraction of solar energy reflected back to space—also affects the greenhouse effect equation. A low albedo number means small amounts of reflection; a high albedo means a large percentage of solar energy reflected. Oceans and rain forests have low albedos, since they absorb a large portion of the sun's light. Deserts, ice and snow, and thick clouds have high albedos, indicative of high reflection properties.[9]

To remember the definitions of high and low albedo, envision a bald "Uncle Al," and a hairy "Uncle Al." The chrome-domed Uncle Al represents high albedo, while the follicular-blessed Uncle Al represents low albedo. In the case of combating global warming, bald is beautiful for Uncle Al.

The albedo effect serves as a positive feedback system. As more snow and ice melts at the poles, more of the sun's energy gets absorbed by the exposed land and sea, which raises the atmospheric temperatures and promotes further melting.

Composition of Air

The atmosphere is composed of a relatively thin layer of gases that sustain life. A typical dry composition of air gases includes nitrogen (78%) and oxygen (21%), along with other small contributors, including carbon dioxide (0.036%).[10]

Notice carbon dioxide (CO_2) makes up a very small percentage of the atmospheric gas. However, it's the one everyone talks about when discussing greenhouse gases which trap heat in the earth's atmosphere. Why? The answer lies in the fact that carbon dioxide is the most plentiful of the partially controllable greenhouse gases and because it can linger in the atmosphere longer than 100 years.[11] (CO_2 atmospheric mean lifetime calculations involve many assumptions and are difficult to estimate. Both sides of the debate use this "uncertainty" to bolster their cause. You'll see later where I'm going with this.)

However, the most plentiful greenhouse gas (GHG) by far is water vapor. So why isn't it the center of GHG discussions? Water vapor makes up more than 95% of all greenhouse gas. The concentration of water vapor changes in volume from 0% to 5% of air due to variations in temperature; nevertheless, humans do little to change the concentrations directly.[12] Here, natural effects rule. In addition, water vapor remains short-lived in the atmosphere, usually only a matter of days. In general, the longer the atmospheric lifespan of a greenhouse gas, the more time it has to work its warming effects. These provide reason for the omission of water vapor in typical greenhouse gas discussions. In fact, the global warming potential (GWP) is not even calculated for water vapor, since it is beyond human control.

Other greenhouse gases, such as methane (CH_4) have an upper atmosphere impact twenty-three times worse than CO_2, but it makes up a miniscule percentage of the atmosphere.[13] However, both natural and human triggers exist for rapid cascading effects that could release massive tonnage of CH_4 to the atmosphere, and quickly. The atmospheric concentration of CH_4 could change quickly if a large "burp" of methane hydrates occurs, which we'll discuss later in this chapter. But first, let's look closer at the poster boy for the greenhouse gas effect: CO_2.

Popping the Cork

"Come quickly, I am tasting stars!"

According to legend, a blind monk named Dom Perignon cried out this phrase the day he first tasted champagne in north-central France. They accidentally discovered champagne when carbon dioxide built up under high pressure in tightly corked bottles of fermenting wine.

We are all familiar with carbonated beverages, from sparkling water to sparkling wine, soft drinks, and beer. Each of these beverages contains carbon dioxide gas trapped under pressure in the bottle or can. When opened, the tiny bubbles of carbon dioxide (CO_2) come rushing to the surface. We also know that beverages left exposed to open air for too long go flat, which means the carbon dioxide has escaped into the atmosphere.

Carbon dioxide remains the poster boy of bad behavior in global warming scenarios, although we know the primary culprit isn't soda or sparkling wine. The CO_2 contribution from humans stem from two main sources: fossil fuels burned to generate electricity and heat, and combustion engines that power the transportation sector of the economy. Similar to the escaped CO_2 from the opened soda containers, the burned fossil fuel releases carbon to the atmosphere in the form of CO_2. The pressure in the container keeps the CO_2 in the liquid drinks. The burning of fossil fuels with air releases carbon into the atmosphere. That carbon, formerly trapped for eons in the earth's crust, is now being released in a geologically short amount of time. The rapid burn rate of the fossil fuels over the last 100 years (especially the last 30) has caused legitimate concerns regarding how fast the earth's natural systems can absorb the additional CO_2 emissions. The concern is that we are adding CO_2 to the atmosphere faster than natural processes can absorb and balance it out.

A Ton of Problems

One of the difficulties in comprehending the magnitude of concern over CO_2 emission is the fact that carbon dioxide is invisible. This makes it difficult for people to visualize the potential problems. Worldwide carbon dioxide emissions reach into the billions of metric tons annually, over 29 billion tons in 2006. In the world, approximately 80 million tons of CO_2 per day spews from smoke stacks, vehicle tailpipes, and flaring of methane.[14] A metric ton equals about 2200 pounds, or the weight of a small car. To understand how many tons of CO_2 emissions come from a coal-fired power plant, consider the following:

- A typical 650MW sub-bituminous coal-fired plant burns about 8265 tons of coal a day, and each coal train-car holds approximately 110 tons.
- Although CO_2 is an invisible gas, the actual weight of the CO_2 leaving the chimney in the same 24-hour period actually weighs more than the weight of the raw coal, by about 3287 tons.

How can that happen? The answer lies in the fact that the oxygen molecules from the air (added in the combustion process) weigh more than the molecules originally attached to the carbon in the coal. The result? Every train car of coal going into a typical coal-fired power plant causes 1.4 train cars worth of CO_2 to exit the stack. Picture that!

Figure 5.2—Coal Train–Power Plant CO_2 Output Illustration. 5 train cars of coal in- 7 train cars of CO_2 out, 104 per day; 3.125 train cars per hour, 4.3 train cars worth of CO_2.

Here's another one. Every gallon of gasoline contains the equivalent of 19 pounds of CO_2 when combusted. Each gallon weighs about 6 pounds. Therefore, the differential between what goes in your car's gas tank, and what comes out the tailpipe is 13 pounds of CO_2, not to mention other undesirable gases.[15] This is possible due to the reduction of oxygen from the atmosphere, combined with the carbon from the gasoline. The resulting man-made (anthropogenic) carbon dioxide would not otherwise exist in nature.

Amazingly, the transportation sector creates the most carbon dioxide in the U.S., with more than 1,900 million metric tons per year ($MMTCO_2$); greater than the industrial sector, which comes in second at 1,672 $MMTCO_2$.[16] Fortunately, the industrial sector can feasibly use large-scale carbon capture and sequestration, although it's expensive. Unfortunately, the transportation sector cannot use this technology. Problems in capturing and containing the CO_2 are more difficult to overcome with millions of vehicles on the road and thousands of aircraft in the air. This makes the search for zero carbon emission alternative fuel vehicles all the more important.

Carbon Conclusion

When you consider that the layer of atmosphere, which sustains life, is relatively thin, (equivalent to the thickness of the skin of an onion in earth scale[17]) you understand that pumping billions of tons of CO_2 into the atmosphere is probably not a good idea. Although CO_2 makes up less than 0.036% (approximately 364 ppm) of the earth's atmosphere, we've come to realize that its effective potential packs a wallop. In comparison, passing a kidney stone the size of a grain of sand can bring a man to his knees.

Although some would say the link between human induced CO_2 and the increase in global temperatures is minor compared to the combination of natural variables, like the release of CO_2 from warming oceans coupled with solar cycles, it still behooves us to do something to reduce our carbon emissions, since we do not have much control over other factors, and the uncertainty of our human impact at least warrants caution. If the human link to CO_2 proves stronger, then our efforts will be all the more important.

Regardless of the strength of the link between climate change and greenhouse gases, the fact is it makes sense to curb our carbon output and pursue cleaner alternatives from other angles, also. Even if you do not buy into the doom and gloom crisis urgency of global warming, remember to stay open to other reasons for change, some of which we haven't addressed yet. Remember that God is in control, but He gives us certain things to shepherd. It's all in the way we use the gifts that He gives us. Recall Genesis 1:28, God's command for us to "control the earth and subdue it." Those words, encompassed in the first sentence of direct command God gave us, remain eternally valid. They should not be taken lightly or handled roughly.

Messy consequences can occur when shaking a bottle of champagne. We need to be smart enough not to put our face in front of the bottle when the cork blows. Likewise, burning hydrocarbons requires caution and constraint, because at the very least, we are unsure of the results. We can reduce our carbon emissions through conservation and efficiency improvements, through application of carbon capture technologies, through reforestation and other carbon sink expansions, and through alternative energy sources that do not release carbon to the atmosphere at all. Carbon dioxide is invisible, but it causes tangible problems.

The solutions, however, come attached with significant cost, leading to the possible cramping of economic growth in developed countries and the constriction of developing countries already squeezed by economic depravity. Some say, "How can we afford to do this?" Others say, "How can we afford not to?" Still others exclaim the cost barriers aren't barriers at all. They claim the perception of barriers exists, but when you include the hidden costs of environmental impacts, the cost barriers disappear, in their view.

Methane

Before we address potential solutions to rapid climate change, let's look at how humans impact the climate without burning fossil fuels. For example, the planting of rice paddies—virtually unknown in the Western Hemisphere—produces a significant amount of methane, a powerful greenhouse gas. The huge growth in Eastern Hemisphere population has added to a greater need of their staple food: rice. People have to eat, and many do not have the economic means to provide anything else as a replacement. Therefore, it's unlikely these greenhouse gas emissions will vanish.

Cattle and other livestock produce considerable amounts of methane out their back ends, but they, too, are part of the food supply for many. On the other hand, if everyone went vegetarian, the longer life expectancy of the animals and their increased consumption of plants would put even more methane (and CO_2) in the air.

Other sources of human-induced methane include leakage from landfills and waste treatment plants, and release from crude oil and natural gas drilling operations.

Methane possesses such a large global warming potential because it breaks down into CO_2. This equals two greenhouse gases for the price of one—not the kind of bargain we need.

Methane Hydrates

Now let's move from CO_2 and examine a form of CH_4, methane-hydrate (also called methane clathrate), often found in deep ocean silt. A methane-hydrate contains a methane molecule trapped inside a lattice of another molecule without connection to it. Think of methane hydrates as a person stuck in a cage. The person isn't part of the cage, but trapped inside. Methane-hydrate is only stable at high pressures and low temperatures. In this way, the cage is a bubble that can burst outward if external pressure is reduced. If an earthquake, landslide, or underwater volcano dislodges an area of the ocean floor high in methane-hydrates, a huge quantity of methane could "burp" to the surface, where the pressure is not strong enough to contain it. Remember, methane is twenty-three times more potent a greenhouse gas than carbon dioxide. This exposes a catastrophic potential shift in greenhouse gases.

Other major sources of methane-hydrate are polar ice and melting of arctic permafrost. The very cold temperature of the age-old ice entraps the methane-hydrates. It is amazing to see chunks of ice burning, but methane-hydrate can do just that when ignited.

Hey Man, Get Off My Cloud

Now let's return to water vapor. Remember that water vapor is a more potent greenhouse gas than CO_2. But since water vapor is such a natural part of the atmospheric cycle, it is pointless to target as a greenhouse gas. Humans cannot noticeably affect water vapor levels. And since water vapor goes on to form clouds, other factors exist.

Clouds can have cooling effects. For example, thick cloud formations reflect the incoming sun. But thick clouds can also hold in heat, acting as a planetary blanket in the evenings. Thin clouds can also increase heating effects because they allow shortwave sunlight through, but trap longer wavelength energy from escaping when it's reflected back from the earth's surface. The net effect of the clouds' heating and cooling balance is a slight, net-cooling effect.[18]

Hurricane Horror Story—Katrina and Rita

Not much good can be said about the horrible hurricanes that decimated the Gulf Coast region, but if nothing else, Katrina and Rita exposed the susceptibility of our oil and gas infrastructure to natural disasters. Long after the ocean subsided, the economic storm continued to surge through our wallets at the gas pump and on utility bills around the nation. Because the world market sets oil and gas prices (setting aside any claims of price manipulation), a strong case exists that Katrina and Rita put a premium on prices worldwide.

Some have gone as far as to claim that we caused, or at least intensified, those storms ourselves. They say we're reaping what we've sown in regard to both our human and economic suffering because of human-induced global warming. Some assert that we have invoked God's wrath. Maybe they're right, maybe they're wrong, but increasing scientific evidence suggests that climate change impacts hurricane development.[19]

Bigger, Stronger, and More Frequent

The Massachusetts Institute of Technology (MIT) and other researchers recently concluded that an upward trend in both the strength and frequency of hurricanes in the Caribbean would continue for the foreseeable future. They believe that the surface waters of the Atlantic Ocean have warmed as a result of atmospheric warming, providing more heat and energy to build and strengthen tropical storms. The storms, birthed off the northwestern coast of Africa, move westward, intensify, and become hurricanes when the winds reach a minimum of 74 mph. A multitude of factors goes into determining if a storm will develop into a hurricane and what course it will take. If we're lucky, straight-line winds from the west blow strong enough in the upper atmosphere to knock down the storm's strength.

Others argue that the storms naturally go through this strengthening cycle about every thirty years as part of the Atlantic Multidecadal Oscillation (AMO). The question remains how much of the variation is natural and how much is intensified by global warming. We're on the upside of the AMO cycle right now, which will make life exciting in the southern coastal regions of the U.S. and the Caribbean.

Regardless of the causes, the impact will be catastrophic. Global population within sixty miles of the coastal regions around the world equals between one-third and two-thirds of the total population, according to some studies, putting many lives in harm's way. Another study that incorporated coastal population with topographic elevation concluded 634 million people are directly in jeopardy because of coastal flooding.[20] Many of the world's largest cities are located near the coast, causing a significant concern. Oceanfront development has grown tremendously, putting property at risk also. And oil production and refining capacity has mushroomed in the Gulf of Mexico region, putting a large part of our energy infrastructure in Hurricane Alley, as Katrina and Rita proved.

The insurance costs and added premiums alone forewarn of the pending disaster. Many insurance carriers are leaving the area because of the high level of risk. Rebuilding in hurricane zones, in storm surge areas, or below sea level, whether you see it as courage or cranial constipation (a dumb move), hurricanes don't care.

Nature vs. Man: Can Nature Absorb the Changes Regardless?

It amazes me how entrenched both the hardcore environmentalists and the hardcore conservatives (for lack of better stereotypes) remain on this issue of natural versus man-made causes. For the extreme camps, the debate rages on as to whether our actions cause climate change, or whether climate change happens as part of a purely natural oscillation in environment. A pretty huge part of the population in the middle is undoubtedly getting tired of hearing about it. No matter whether you attribute the ability of the earth to regulate the environment to Mother Nature or to God, everyone can agree that the climate has changed throughout history and that built-in natural mechanisms react to changes in the atmosphere and oceans. For example, every breath that humans exhale creates carbon dioxide. Plants use the CO_2, and expel oxygen, which we need to breathe. The elegant balance of this design is wonderful.

For larger scale atmosphere imbalances, if there's too much carbon dioxide in the air, oceans can absorb carbon dioxide, forming carbonic acid, the same effect as the formation of soda. Absorption of CO_2, called a "carbon sink," remains nature's way of balancing atmospheric CO_2. Oceans ingest the largest percentage of CO_2, mainly through the growth of plankton in the upper regions of the ocean. The problems that occur concern how quickly they can react to balance out an atmospheric change and what effect it has on the ocean environment. Increases in the ocean's surface absorption of carbon dioxide can increase the acidity of the ocean,[21] which contributes to *coral bleaching*, which effectively kills coral reefs. The beautiful coral reefs support abundant forms of ocean life. Unfortunately they grow slowly and cannot adjust to ocean changes, whether it's a rise in CO_2 or increasing ocean temperatures.

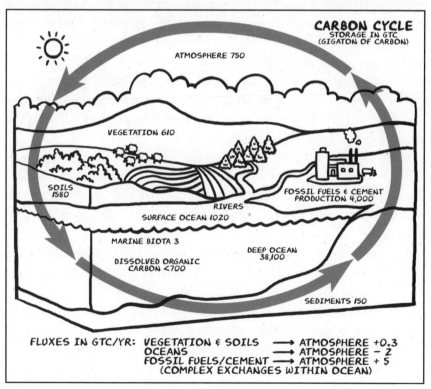

Figure 5.3—Diagram of Carbonate Cycle.

The other natural carbon sink, second only to the ocean, is forestland, especially rainforests. They absorb tremendous amounts of carbon dioxide, especially in young new-growth forests, but also have a negative feedback due to their low albedo, which traps heat and humidity. But the greater good by far is the increased carbon dioxide absorption. Unfortunately, the forest can also become a natural liability because forest fires quickly release all the stored carbon back into the atmosphere. This is somewhat tempered by the effect of the smoke that's produced (which lowers the atmospheric temperature by blocking sunlight), but the benefits of blocked solar radiation do not last as long as the problems caused by an increase in carbon dioxide. Overall, forests are part of the solution for nixing carbon dioxide increases.

We've talked about nature's mechanisms for reducing CO_2, but natural increases are also quite possible and sometimes dramatic. Volcanoes are often identified as a primary culprit, but in reality, their effects in modern times are likely the opposite. Although volcanic eruptions spew over 100 million tons of carbon dioxide and methane into the air on average per year, that total is less than $1/100^{th}$ of the human-generated contribution. However, unlike forest fires, ash from volcanoes can have extensive cooling effects based on its properties (primarily sulfur content) and the height to which it expels the gas and ash into the atmosphere or even the higher troposphere.

The higher the ash is propelled into the upper-atmosphere, the longer it blocks out the sun's rays. This reflects significant amounts of the sun's energy, which has a cooling effect, but results in reduced plant growth and mortality, which indirectly increases CO_2. The net effect is one of cooling.

The composition of the volcanic gases and ash is even more important than the size and force of the explosion. In recent history, volcanoes tell an interesting story.

The 1980 eruption of Mt. St. Helens ejected tons of ash into the atmosphere, which resulted in lower global temperatures of approximately 0.1 degree C.[22] The much smaller eruption of El Chichon in Mexico in 1982 resulted in three to five times the temperature decrease because of the higher sulfur content of the El Chichon eruption—about forty times greater. The strongest predicator to climate change from volcanoes appears to be gas composition, not the volume of ash. The high sulfur gas combines with water vapor to form miniscule sulfuric acid droplets that can take years to settle out of the troposphere, and the droplets reflect solar radiation back out into space. In 1991, Mt. Pinatubo in the Philippines produced even more airborne sulfur than El Chichon, resulting in the largest sulfur oxide emission of this century. Mt. Hudson in Chile erupted the same year,[23] resulting in a rare downturn in temperature from the upward march of global warming from 1989 to present.[24] Volcanic eruptions remind us that no matter how hard we try, we will never be able to completely predict, much less control, all aspects of our environment.

Sun, Wobble, and Elliptical Variation

We cannot control the intensity of the sunlight reaching the earth, of course. Three major items not directly related to greenhouse gases affect this, but they do affect earth's temperature: the sunspot cycle, the earth's axial wobble, and the earth's orbit.

The sun goes through a cycle about every eleven years where the energy output of the sun's fusion reactions increase and decrease. It's called the sunspot cycle. The sun and other stars suck in huge amounts of hydrogen from their solar systems. Through the intense heat reactions, the hydrogen nuclei are fused together to form helium atoms. Only a small fraction of the sun's energy is absorbed by the surface of the earth, and the average is 168 watts per square meter.[25] This fairly constant average does vary enough to make a difference felt. At times of intense sunspots, the intensity of light falling on the earth increases, boosting the average watts per square meter and working to raise the global temperatures. We can measure and calculate how much average temperatures rise, but we cannot completely discern the amount attributed to sunspots or any other cause.

The earth has a wobble to its axial rotation which also impacts climate change. The variation in this axis wobble can be as much as twelve degrees. This tilt can affect the amount of light the more heavily vegetated Northern hemisphere receives, impacting the growing season and CO_2 levels.

The earth's elliptical orbit also has an impact. The shape of the orbit itself, and the fact that the shape of the orbit also varies cyclically, impacts the earth's climate temperatures.

Of course, we cannot control these items, but we must be aware of them to understand the situation. "Acts of God" that He sets forth in nature will always trump our human efforts, but that does not give humankind a blank check to avoid responsibility.

Climate Chaos

So who's really to blame for global warming? What if humankind is the primary cause of the recent increase of global temperatures? What if our continued addiction to fossil fuels and exploding population growth really is to blame?

On the other hand, throughout the earth's history, the world has warmed and cooled due to natural variations in the sun, variations in the earth's orbit, volcanic eruptions, and changing of ocean currents. We can control none of this. So should we alter our lifestyles and spend billions of dollars to try to stop it? Is it a wasted effort to try and control climate change?

That leads to another question. What will result if global warming continues?

We have many unanswered questions, which we may never be able to answer scientifically. We cannot determine today the mechanisms that control climate, and worse yet, we may not be able to do so in the future. But why?

To explain, let's go back to the 1970s. Scientists spawned an idea, attempting to explain why certain systems, like the weather and climate change, remained beyond the scientific grasp of accurate prediction. This became known as "chaos theory." Basically, chaos theory assumes that some systems have so many random variables that scientists would forever be limited in their ability to accurately predict the future. Since then, computer models have become more powerful, allowing us to better predict weather and climate changes, but chaos theory remains stubbornly true. Thousands of variables can affect climate change. Few of the variables can be quantified, and we don't even know some of the variables. We will never be able to accurately determine the exact mechanisms of global climate change. We can measure temperatures and a variety of probable causes to hypothesize the causes of climate change, but we cannot definitively pinpoint all the causes and effects, or dole out percentages of guilt. The best we'll likely ever be able to do is predict by probabilities and general cycles.

Weather forecasting brings us a perfect example of chaos theory in action. No offense to the thousands of meteorologists out there, but how many times does the weather forecast fail to meet prediction? Certain regions are harder to predict than others, but everyone has experienced an incorrect weather forecast. Weather forecasts often fail expectation because they are a compilation of ever-changing variables, such as temperature, moisture, wind speed, ocean temperatures and other weather systems. Those variables are affected by dozens, if not hundreds, of sub-variables. It is chaos coming up with an accurate forecast. Since we cannot predict accurately, we definitely cannot control weather or climate.

Chaos, the Cat, and the Kid

Picture a huge universal remote control with 2 million buttons on it. That shouldn't be hard—my TV remote has about 1.5 million right now (or at least it seems like it).

Figure 5.4—Chaos Theory TV Remote Illustration.

Now, imagine that 1 million of those buttons have no labels. Now picture a cat and a two-year-old continuously walking on the buttons. That illustrates chaos theory.

The climate, like our remote control, features certain buttons with more effect than others. Unfortunately, with climate change, we may not have the time to learn the remote. For climate change, as with a kid's cartoon movie, the show might be over before we learn how to operate the remote.

That's why it's not as important to fix blame on Mother Nature or human nature in regard to climate changes. If the cat and the kid represent nature and humankind respectively, and they both mess with the remote simultaneously, then both affect the TV, but it's hard to tell who's doing exactly what. Humans certainly affect climate, but many still prefer to "blame it on the cat."

Global Dimming

Another factor in the *nature vs. man* debate on the causes of global warming is the phenomenon called global dimming. Of all the research and discussions between sides of the climate debate, this could be the missing link that explains why all the fossil fuel burning that we've done over the last century has not caused a larger increase in global temperatures.

The global dimming theory is based on the idea that condensation forms on the fine pollution particles from power plants and internal combustion engines. These particles reflect a lot of sunlight back into outer space and disrupt the natural formation of water droplets. The cooling effect that results slows the rate of global warming.

According to a *Nova* special first aired in April 2006,[26] the effects of dirty air versus clean air are most pronounced in the Eastern Hemisphere, where coal-fired power plants in the developing world have grown tremendously without the pollution "scrubbing" technologies mandated in the Western world. This results in a far-reaching band of pollution over the Indian Ocean that (via the jet stream) reaches the southern Maldives. The air to the south of that jet stream is pristine in comparison. By comparing the atmospheric conditions from north of the jet stream to south of the jet stream, it's obvious that a reduced amount of solar radiation reaches the earth's surface. In effect, the pollution acts as a shield, rejecting a portion of the sun's warming energy before it penetrates the atmosphere.

The tug-of-war between greenhouse gas warming effects and the cooling effects of global dimming nearly balance, with a lean to the warming side. The global dimming phenomenon accounts for the less-than-expected increase in global warming. Is the answer then to roll-back the pollution mitigation legislation the West has enacted over the last three decades? No. Taking backwards steps solves nothing. Perhaps a better answer is to reduce pollutants in the developing world and suppress carbon emissions in both the developing and developed world.

Global dimming theory could go a long way in explaining past differences between predicted versus observed results of climate models. The failure of past climate models to predict the magnitude of climate change, which was often overstated, provided considerable fodder to skeptics of climate crisis. Conspiracy theories exist, often claiming that environmental demise exaggeration results from scientists in the pocket of the environmental lobby, keeping research dollars flowing by deception.

Michael Crichton took this approach in his book, *State of Fear*, where extremist environmentalists not only exaggerated claims but actually triggered natural disasters to further their agendas. Although purely fictional, it included a considerable amount of technical information. It caused skepticism toward the dangers (or reality) of climate change among those who read the book. Unfortunately, this is the only exposure to climate science that some have, and the fictional format probably isn't the best presentation for a complex subject. It is good practice to "test everything," as it says in the Bible. Hopefully, global warming skepticism and fanaticism will both result in deeper investigation. Theories such as global dimming might open some dialogue. It can explain why the warming effect is not as strong as theorized, such as during the recent global cooling period from the pollution and sulfur dioxide-ridden years from around 1940 to the early 1970s.[27]

The Unknown Balance

So what is mankind's impact on the global climate? Compared to some of the naturally occurring phenomena, it might be tempting to justify our actions in burning fossil fuels as inconsequential, or so diffi-cult to correct and control that it is a WOTBAM (waste of time, brains, and money). However, many things worth doing are difficult. And if it is worth doing, it is worth doing right, as the old adage claims. In fact, a lot of money could be made in environmental cleanup technologies, providing good-paying jobs. Considering the state of the economy in 2008 and 2009, many are connecting with the idea of "green jobs" that could grow new infrastructure and revive the economy. Perhaps the smartest action is not to fight the clean technologies but to embrace them.

The combustion of fossil fuels is overwhelmingly number one on the human impact scale. The production of gasoline and other distilled fu-els, from "well-to-wheels," harms the environment. Recently, in spring 2006, transportation-sector burning of fossil fuels took over the top spot for generation of carbon dioxide in the United States. Electrical power generation runs a close second. The remainder comes from the industrial sector and residential heating. It's almost impossible to cap-ture CO_2 from transportation emissions, therefore, until all-electric or hydrogen vehicles hit the market in force, it will be difficult to reduce transportation-sector emissions.

Focusing on the electricity generation sector, it is possible (but not economically feasible) to capture and *sequester* the carbon dioxide. *Sequestration* pumps the carbon dioxide into underground cavities capable of being sealed, thereby trapping the CO_2 underground instead of releasing it to the atmosphere. A potentially favorable marriage of technologies enhances oil recovery from depleted wells by pumping the carbon dioxide into the well. The CO_2 injected in the well reservoir increases its viscosity and the added pressure forces oil to the surface that would otherwise remain un-recovered. This is one method of secondary recovery. The additional benefit is the capture of the carbon dioxide.

Carbon sequestration has just begun large-scale implementation outside the United States. Depleted natural gas reservoirs, coal bed methane fields, and stable, deep ocean locations could also be used. We need more experience and research to verify the safety of these ideas, but they appear promising. Even if not a perfect solution, they promise better results than the alternative release of CO_2 to the atmosphere.

Kyoto: Conform or Cut Bait?

If you want to start an argument in a room full of conservatives and progressives, just say one word—Kyoto. The Kyoto Protocol seeks to stabilize internationally greenhouse gas emissions from a number of sources, but fossil fuels are a key player. In particular, fighting fossil fuel emissions from coal seems daunting. To the dismay of hardcore environmentalists, it's not practical for coal to disappear overnight. Nevertheless, the business-as-usual approach to coal is equally impractical—just try to get an air permit for a coal plant.

However, it's important not to forget that the various revisions of the Clean Air Act in the United States have greatly improved air quality by reducing sulfur dioxide, nitrous oxide, and particulate emissions from power plants and other industries. Mercury has also been added to the list, but carbon dioxide reduction remains a voluntary initiative in the United States.

Another air improvement success to celebrate has been the Montreal Protocol. It internationally restricted chlorofluorocarbons (CFCs), strictly man-made gases (mainly refrigerants and aerosol propellants) which deplete the upper-atmosphere ozone layer and also warm global temperatures. We often forget about these improvements which changed the business-as-usual case at an increased financial cost at the time, but are now part of the norm. Will we some day take carbon restriction for granted the same way?

The international attempt at CO_2 reduction is tied to the Kyoto Protocol, named after the Kyoto, Japan, location where the summit meeting was held in 1997. The Kyoto protocol is based on a cap-and-trade system that uses the year 1990 as a benchmark. Simply, the carbon dioxide emissions of a given country that has signed the protocol are set at the 1990 level and they must reduce that by 5% between marker years 2008–2012. If the countries do not meet their emissions "cap," they can "trade," for a price, with another nation which is significantly under their cap.

When Russia signed the treaty in 2005, the percentage requirements to ratify the treaty as a whole were met. Australia ratified the treaty in late 2007, after a political change in power. The United States symbolically signed during the Clinton administration, but congress made no serious attempt at ratifying it for almost a decade. The paramount reason not to sign involves a combination of economics, politics, and concerns over fairness and effectiveness; namely, emerging industrial giants China and India are not included under mandatory limits. The rationale for this and for other developing nations is that they had minimal historical greenhouse gas (GHG) emissions and their fledgling economies would collapse under the weight of Kyoto.

This puts the United States and other industrialized countries at a big disadvantage if they commit to Kyoto, with estimates for compliance in the billions of dollars. Major concerns of economic hardship are now heightened by the economic collapse, coupled with little guarantee of effectiveness or net benefit of GHG reductions in return for ratifying Kyoto. This becomes especially true if developing countries, such as China and India, remain exempt from the next stage of climate treaties and are not held to the same standard.

Right or wrong, the United States has chosen not to participate and instead established a voluntary plan for businesses to reduce their greenhouse gas emissions. President Obama and the Democratic congress may change course on accepting international carbon protocols, pending the health of an economic recovery, but the core question remains of how to make carbon reduction effective going forward in the future. Kyoto, which is Phase I of a multi-phase program, ends in 2012. Regardless of its symbolic importance, Kyoto has little chance of reaching its intended overall goal. It taught us many lessons to carry forward and has provided a valuable impetus to create carbon trading markets and promote renewable energy, but nonetheless falls short in many areas.

However, the global and political ramifications of America's detachment make it difficult for the advancement of worldwide environmental progress. Without our involvement, any international climate treaty chances at success are probably dead on arrival. Our global leadership and reputation are at stake here. It weakens our nation significantly if history proves us wrong on Kyoto and the timing of our commitments. If nothing else, we missed many opportunities to enact something better or explain our position better. To many in the world community, we appear selfish because of our inaction on Kyoto.

Regardless, over ten years have passed since Kyoto and little effective change has occurred. We can continue to play "what if" regarding the Bush presidency, assess blame to this or that political party, or lament the flawed shortcomings of the Kyoto treaty, but that will do little to change the future. Far too much time, brains, and money have been poured into ineffective squabbling over Kyoto. The answer is to find something more effective, whether it's an improved second-stage protocol that provides hope of better balanced results, or a grass-roots change in how we supply power to transportation and electrical segments of the global economy.

"So What?"

What will happen to the planet if global temperatures continue to rise? Like standing on the volume button on that remote control we discussed, eventually the speakers will blow. The same holds true for the atmosphere. We need to turn down the carbon volume, to at least attempt to shift to sustainable alternative energy and to help the rapidly shifting climates. It's important to remember that there's more to the story than just temperature; there's also precipitation and clean water.

Patterns of flood and drought can intensify as a result of our actions. Likewise, the added world population increases the burdens felt from climate changes. The added stress to the environment affects clean water supplies, too. One thing few realize is that clean drinking water is at heart an energy issue. Over two-thirds of the earth's surface is covered by water; energy is required to purify it and pump it where needed.

Doing nothing risks destroying the showcase of God's glory—the beautiful earth He graced us with as a gift. In the first chapter of Genesis, God gave man responsibility via His command to "fill the earth and subdue it" (Gen. 1:28). The word "subdue" does not justify abuse of the earth. Nor does "subdue" imply we should inhibit our use of the earth. Subdue means to calm and tame the earth, not destroy it.

So far the answers to the questions are depressing or non-existent. We cannot accurately determine what drives climate change; we cannot effectively control it; and we don't know with certainty how climate change will affect the future. We cannot even agree on the primary cause, human or natural.

CALM IN THE FACE OF CHAOS

It's time for the bombshell. By and large, we're asking the wrong questions. We should ask questions like these:

- Is it a good idea for humans to make questionable contributions to global climate change?
- From a stewardship perspective, what can I do about it?

The answer to the first question is no. Humankind has a horrible track record when it comes to caring for the environment. We tend to act without considering the results of our actions. Pursuit of cheap energy is fine to a point, but we must be careful not to overdraw the "cheap energy" bank account at the expense of the environment.

Medical doctors take a vow to do no harm. We need the same creed for our environment. Because recent climate changes have happened so rapidly, not allowing nature to take its course as God intended, we must now attempt to rein in rampant climate change, regardless of its cause. Even if CO_2 is not the primary driver of recent global warming (and most would argue that it is), the stakes are too high to sit idly by.

Billions more humans live on the earth at this time than in previous times of major climate change. As Christians, can we say that poor treatment of the earth doesn't matter? On the contrary, God has tasked us with being good stewards of the environment, including the atmosphere and environment. He calls us to love our fellow man and care for his physical needs, in addition to his spiritual needs. A healthy atmosphere, low in pollution and detrimental effects, is critical to Christian caring. How will others perceive the actions of Christians in regards to the environment? Will it hinder our mission of caring for others spiritually if they witness our indifferent or harmful actions to the environment? Christians should strive to care for the environment, to do no harm, and prove our love of God's creation by treating the earth with respect.

The second question is personal. We must all realize that environmental stewardship isn't just the responsibility of countries and companies. Countries and companies are entities. These entities are an "it," not a "they." Companies do not read books, and countries do not burn hydrocarbons. The people who make up these entities are collectively responsible. That means all of us, from CEO, to stockholder, to politician, to citizen of any industrialized or developing country.

A case in point: In the United States, petroleum makes a large contribution to global warming gases, approximately 44% of CO_2 emissions in 2006.[28] Does that make it a sin to drive a large SUV, truck, or van? Not necessarily, and we should all be cautious not to rush to judgment since there is plenty of blame to go around. If the SUV transports a large family around, or if the truck regularly hauls payloads, then it serves a useful purpose. However, a better choice likely exists, and more and more Christians are asking W.W.J.D. (What would Jesus drive)?

It does not mean we should trade our SUVs for mopeds tomorrow. We can only sustain rational choices. But the SUV owner could conserve fuel by driving slower, carpooling, keeping the tires properly inflated, or keeping the air filter clean. Maybe they should consider purchasing a hybrid or a bio-diesel powered model a little sooner than they planned, although that is also a transitional technology.

The bottom line is not to focus on the blame or wash our hands of the situation because of the uncertainty. Maybe we don't fully understand climate change. Maybe we don't have any hope of controlling it, or truly know what the future impacts of climate change will bring. However, we should continue trying to do the right thing on a personal level. We can control our response to climate change in a moral fashion, even if we cannot control the climate itself.

Perhaps we should hand over the remote to God and ask Him to fix it. Better yet, we should pray to Him to show us which buttons to hit, for His glory.

Thankfully, there is hope in the answer to those prayers. Scientists on both sides of the "climate doom" versus "climate contrarian" arguments agree that globally the climate is changing, although they disagree about the cause of the climate change and the future. Regardless of who is most correct, I am convinced there is hope in the resiliency God has instilled in the earth. We can turn the ship around, without destroying the economy of industrialized nations and without destroying middle-class growth in developing nations. But it won't be easy.

If we want to be better stewards of our environment, we have to move to sustainable energy. I firmly believe that if you want to change the existing reality, don't just attack the conventional wisdom; create a new path that reveals the treachery of the old path. That's where hydrogen and fuel cells come in.

CHAPTER 6

HYDROGEN, FUEL CELLS, AND SANTA CLAUS

> All conservatism is based upon the idea that if you leave things alone you leave them as they are. But you do not. If you leave a thing alone you leave it to a torrent of change.[1]
> —G. K. Chesterton

Do you believe in Santa Claus? You cannot deny the fact that he's an economic juggernaut. Hydrogen and those mysterious things called fuel cells that we'll learn about later may not yet hold this distinction, but I believe they will. However, many unbelievers claim the future of hydrogen and fuel cells is as much a fantasy as Santa Claus. Some have gone so far as to label them "fool cells." They focus on all the hurdles that hydrogen and fuel cells must overcome to succeed: from technical, to economic, to public acceptance. After all, hydrogen has been around since the beginning of time; why should we change how we use it now? The question is—should we believe in the potential of hydrogen and fuel cells? Some would say you might as well believe in Santa Claus.

Well, hold on to your reindeer: it turns out that Santa Claus was a real person, Saint Nicholas of Myra, a 4th-century Christian known for his knack of generous, anonymous giving. The name Saint Nicholas has morphed over the years and across cultures to include the American mispronunciation "Santa Claus" from the Dutch name "Sinterklaas," which is a contraction of "Sint Nicolaas." As a wealthy and devout Christian, St. Nicholas applied his God-given gifts of wealth and social position wisely to help the poor in the region of Myra, located in present-day Turkey. Of the legendary stories told, one of the more famous anonymous gifts credited to him is the dowry money given to a poor family so that the father's three daughters could marry and would not have to enter the world of prostitution to survive. The gift of hydrogen and fuel cell technology has similar benefits: with its adoption, we will be able to rise up to a higher form of energy usage, rather than being enslaved by less desirable forms of energy necessitated by survival.

St. Nicholas went from an obscure saint to one of the more recognizable figures in modern history. Will the use of hydrogen as a fuel in fuel cells make a similar journey? Will hydrogen be recognized as *the* premier fuel of modern history? What about those expensive but promising fuel cell technologies we hear about on the news? Will they ever succeed? Before we learn more about fuel cells, we should ask first, "What is hydrogen?"

Hydrogen is #1

Hydrogen gas (H_2), which holds the first spot on the periodic table, is the most abundant element in the universe. In fact, it's the fuel for the sun. But on earth, all the pure hydrogen gas escaped the atmosphere long ago. Although plentiful hydrogen atoms remain, they're always joined with another element, such as carbon or oxygen. The most famous example of this joining is H_2O, or water. In fact, hydrogen's root name is "water-former" from the Greek root words *hydro,* meaning "water," and *gen,* meaning "former." The German word for hydrogen, wasserstoff, is even more direct: "the stuff of water."[2] When you consider the volume of water in the earth's oceans, lakes, and rivers, it's difficult to imagine ever running out of hydrogen. We would be more likely to run out of air.

The volume of water available underscores a key advantage of using hydrogen. Another advantage of hydrogen fuel cells or burning hydrogen: water vapor is the only result. But in order to enjoy the benefits of hydrogen, it must be separated from its adjoined elements, such as oxygen (in an H_2O molecule of water) or carbon (in a CH_4

molecule of natural gas). And that's difficult, because as we discussed earlier, hydrogen is a "joiner."

Therein lies the dilemma of hydrogen. First, we require energy *input* to break the hydrogen free from its adjoined atoms, and that detracts greatly from the gross energy *output* we hope to gain from the hydrogen itself. Secondly, once we have the hydrogen pure, and free of other elements, it is such a small and active molecule that it is difficult to bottle-up and store. These significant obstacles, while challenging, can be addressed. That's why I believe hydrogen and fuel cells will be the primary source for our future energy demands . . . eventually. Eventually might mean two years from now, ten years from now, or fifty years from now, but hydrogen will eventually become number one on more than just the periodic chart. I cannot help but wonder if hydrogen's path from obscurity to stardom will be similar to St. Nicholas'. If so, the legacy that hydrogen leaves—the modern gift for mankind—will be hydrogen fuel cells.

Fuel Cells 101

The first fuel cells for a vehicle were developed in 1959 to power a modified farm tractor, but the technology didn't really begin to gain prominence until used in the U.S. space program missions of the 1960s. Today, the most common type of fuel cell, one under rapid development for vehicular transportation uses, is the Polymer Electrolyte Membrane (PEM) fuel cell, sometimes also called a Proton Exchange Membrane fuel cell. A PEM fuel cell produces DC electricity (direct current electrons) by converting chemical energy from hydrogen and recombining the hydrogen with oxygen at the end of the process.[3] Combustion does not occur and the cells are stacked in electrical series and parallel combinations to provide the desired voltage and current.

I like to think of it as a sandwich. Just as you need a few basic elements to create a sandwich (bread, meat, and condiments), a fuel cell also contains basic elements: electrodes, catalysts, and electrolytes.

Electrodes come in sets of two, like the bread of the sandwich: one cathode and one anode. Fuel is introduced to the anode. At the opposite electrode, oxygen is usually provided by exposing the cathode to air. Electrodes have a few purposes:[4]

97

- They provide a path for the external current to flow to and from the fuel cell.
- They provide the interface zones between the electrolyte and the fuel, and between the electrolyte and the air (oxygen).
- They sometimes provide physical support to the electrolyte material.

Catalysts are like the mayo and mustard layers of the sandwich. They help speed the reaction along or lower the temperature at which the reaction occurs. They increase the rate of the reaction by lowering the activation energy required without being consumed. In other words, they facilitate and speed up the reaction between hydrogen and the electrolyte at the anode and between oxygen and the electrolyte at the cathode.

Electrolytes are like the meat of the fuel cell sandwich. They conduct ions from one electrode to the other. The ions that travel through the electrolyte can have either a positive or negative charge, depending on the fuel cell technology type. Fuel cells receive their name from their type of electrolyte.

Our fuel cell stack, or sandwich, is done. If you don't like sandwiches, some types have a tubular design, more like a wrap.

See the simplified diagram below for the basic fuel cell operation.

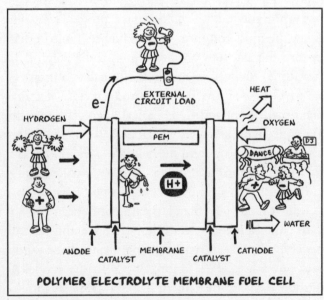

Figure 6.1—Simplified Fuel Cell—High School Love Story Illustration.

The Teenage Love Story

After I acquired a small demonstration model fuel cell for fun, I started doing presentations at schools to explain fuel cells to others. I needed a relatable visual analogy for how the fuel cell process works. That led to a little skit called "The Teenage Love Story."

Basically, the story takes the main components of a fuel cell—the electrodes, the catalysts, and the electrolyte—and assigns them to characters or to scenery set structures. The scene and starring roles are:

- Harry "The Proton" Hydrogen, Male Lead
- Electra Electron, Female Lead
- Electrodes, The Front and Rear Doors of the School's Locker Room
- Electrolyte, The Boys' Locker Room
- Catalysts, Bullies, and Towel Snappers in the Locker Room
- External Circuit, Electra's Hair Dryer and Cord
- Oscar Oxygen, The DJ and His Crew at the School Gym Dance
- Water, The Happy Ending, the Friends Get Together at the Dance

SCENE 1—THE ATTRACTION

The story begins with our young couple, Harry and Electra, at the end of a high school football game. Harry "The Proton" Hydrogen and his beautiful cheerleader girlfriend, Electra, are attracted to each other, like any proton and electron should be. They go together from the field to the school, where Harry asks Electra to go to the dance that night at the school gym. Oscar Oxygen, a popular DJ that both of them really like, is playing at the dance. She accepts. But when they reach the boys' locker room front door (anode electrode), Electra is not allowed to enter because she is a girl (negative charge). Harry (positive charge) needs to clean up from the football game and Electra needs to go home and get ready, redoing her hair with the hair dryer (external circuit). So they agree to temporarily split up and meet later at the dance.

SCENE 2—THE SEPARATION

Electra hurries home along the path (external electrical circuit to the hair dryer) to get ready. Meanwhile, Harry enters the boys' locker room front door and starts to get snapped with towels immediately by the rowdy boys at the front entrance (anode catalyst). They chase him into the locker room (electrolyte). This towel snapping causes Harry get ready quickly (the catalyst is doing its job) and head for the rear exit of the locker room door (cathode electrode) that leads to the gym where the dance has begun. At the same time, Electra finishes with her hair dryer (external circuit) and she hurries toward the school gym. Harry finishes in the locker room and, as he runs to the rear exit, eyes another group of towel snappers (cathode catalyst) waiting for him. He lowers his head and blasts through the throng, bursting through the rear doors (cathode electrode) and into the gym.

SCENE 3—THE BIG DANCE

Harry and Electra enter the school gym at the same instant. They see each other across the crowded room of other protons and electrons and quickly join together. Other protons and electrons are likewise joined and they all dance to Oscar (oxygen) and his band of DJs until everyone is sweaty (water) from dancing. The end.

Hopefully this helps explain the operation of a fuel cell. If not, you'll probably never look at a young high school couple the same way again. (Sorry.) But the point to remember about hydrogen fuel cell technology is that it's not a new concept and it's not particularly complicated. Even better, we have many ways to generate hydrogen. We don't need to rely on any one hydrogen-generation technology to succeed, since marketplace and competition will give us ample supply and the best price, if we don't mess it up. Looking at the number of options is almost as exciting as watching a child survey his presents on Christmas morning.

The Gift of Hydrogen

Just as Santa doesn't bring all the gifts at Christmas (we also receive gifts from parents, siblings, and well-meaning aunts), hydrogen can come from a multitude of sources, such as:

- Wind electrolysis
- Solar electrolysis
- Off-peak, base-loaded power plant electrolysis (usually nuclear or coal)
- Steam Methane Reforming (SMR)
- Coal gasification
- Biomass gasification
- Anaerobic digestion
- Thermal-cracking of water (by high temp solar-thermal, Gen IV nuclear)
- Biological (produced from algae or bacteria)
- Photo-chemical (similar to solar).

Some of these are green sources of hydrogen, which means it's derived from a renewable source like solar or wind. But if the hydrogen comes from fossil fuel powered electricity or from stripping it from a fossil fuel, it's considered a black source of hydrogen.

One of the most common means of generating hydrogen is a process called electrolysis, where you isolate a hydrogen molecule by running DC electric current through water. In its simplest form—which you may remember from your high school chemistry lab—all you need is a glass of water, a battery, and two wires.

ELECTROLYSIS:

$$\text{Electricity (DC)} + 2H_2O \xrightarrow{\text{heat}} O_2 + 2H_2$$

The electrolysis process breaks up the hydrogen bonds that hold the H_2O (water) molecule together by providing more energy than the H_2O molecule bonding energy. The hydrogen atoms and oxygen atoms disassociate to their elemental gaseous forms and are collected in separate containers: hydrogen at the negative electrode and oxygen at the positive electrode. Any technology used for generating electricity, like wind and solar, can be used with electrolysis to generate hydrogen. Solar-thermal energy can also be used to generate sufficient temperatures to drive or assist thermal-chemical hydro-cracking (water splitting).

But the cheapest, most common way to generate hydrogen today, especially in large quantities for industrial use, is steam methane reforming (SMR), a process that involves methane gas. Methane is the largest percentage component of natural gas and has four hydrogen atoms for every carbon atom; its chemical formula is CH_4. In the traditional SMR process, an external source of heat is used to heat water and generate steam. The steam is then reacted with methane gas in the presence of a catalyst to create hydrogen and carbon monoxide (CO) as seen in step 1 below. The carbon monoxide reacts with more steam, causing the water-gas shift reaction to occur in step 2.[5]

$$\text{Heat}$$
$$\text{Step 1: } CH_4 + H_2O \rightarrow CO + 3H_2 \text{ (Called synthesis gas or "syngas")}$$

$$\text{Heat}$$
$$\text{Step 2: } CO + H_2O \rightarrow CO_2 + H_2 \text{ (Called bio gas)}$$

The overall methane reforming reaction with a nickel metal catalyst and heat at approximately 1500 degrees Fahrenheit results in the reaction below:[6]

$$CH_4 + 2H_2O \rightarrow CO_2 + 4H_2$$

The efficiency of traditional SMR is about 80% in large-scale operations and new processes promise to do it better and cheaper. Sometimes part of the natural gas feedstock is burned to create the heat to drive the step 1 and step 2 reactions above. This is called Auto Thermal Reforming (ATR) and the partial oxidation-reduction that occurs is a hybrid method of SMR. For partial oxidation to occur, part of the gas burns, but just a small portion because the majority of the natural gas is needed as CH_4 feedstock in step 1 above. Small scale SMR is usually ATR type technology and on this scale, performs at around 55% effeciency.[7]

Unfortunately, SMR also generates carbon dioxide and most of this carbon dioxide is released to the atmosphere. Therefore, typical SMR would be considered a "black" hydrogen process. But if the carbon dioxide were captured and sequestered (injected into underground caverns, deep into the ocean, or by reforestation), the intended byproduct (hydrogen) could be considered green or at least dark green.

Capturing and sequestering carbon dioxide is an important component of any environmentally friendly equation for any fossil-fuel-derived hydrogen. Because most people associate the word *sequestration* with long jury trials, let us be clear that CO_2 sequestration means retaining CO_2 in seclusion after separating and capturing it from its fossil fuel victim. For fossil-fuel-derived hydrogen to work in a hydrogen economy, it needs the dark green pedigree of sequestration.

Santa's Bad Boy Fuel

What does Santa bring to bad little boys and girls? That's right, a lump of coal. And no little boy or girl wants to receive a lump of coal. But maybe we can have some fun with coal after all. Like all fossil fuels, coal contains hydrogen, and coal gasification holds promise as a way to release that hydrogen before we burn the coal. Coal is a natural resource found abundantly in many large, energy-consuming nations, such as the United States, China, and India. The problem with coal— past and present—continues to be its cleanliness. It's a dirty job to mine it, process it, and burn it. Setting aside the mining problems at the moment, many of the emissions problems with coal have been reduced by clean coal technologies in recent years. These new technologies are expensive to implement, but technically feasible and improving. Removing particulates (such as sulfur dioxide, nitrous oxides, and most recently, mercury) from power plant flue gas greatly improves air quality. However, the biggest and most expensive problem is what to do with carbon dioxide, or CO_2.

In the United States, coal-fired power plants are one of the cheapest ways to generate electricity. But adding CO_2 capture (post combustion) to a traditional coal plant could add 40% or more to the consumer cost of electricity, due to the additional equipment required and the huge quantity of CO_2 in the flue gas.* CO_2 comprises about 14% of the flue gas that's emitted from the chimneystack which, for a 600MW

---

* We could use some of that Santa Claus magic in cleaning up our chimney problems. Only Santa Claus can get up and down soot-filled chimneys in a bright red and white suit and not get a smudge on him.

power plant, equates to about 1 million pounds per hour. If the CO_2 is captured and sequestered, it's much better environmentally—but only to a point, because more energy and, therefore, more coal is needed to operate the CO_2 capturing and sequestering equipment, which reduces the net benefit of the CO_2 capture and containment.

Luckily, a better method of CO_2 capture exists than post-combustion capture. Hydrogen from coal gasification can be considered much closer to green hydrogen because it is easier to remove CO_2 from a gas stream before combustion than after. That is the biggest potential benefit of coal gasification, even beyond generating hydrogen. This is where coal can have a beneficial impact in transitioning to a hydrogen economy.

The most common uses of coal gasification technology are for generation of industrial gases, but there is increasing interest in using coal syngas (H_2 + CO) in Integrated Gasification Combined Cycle (IGCC) projects for electrical power generation. At the moment, IGCC projects do not capture and sequester the CO_2 because of the additional cost. However, the FutureGen project and others hope to prove the viability of these additional steps.[8] Rather than burning the resulting hydrogen-laden coal syngas in IGCC plants, hydrogen's future lies in separation and use in fuel cells, which is another component of the FutureGen project. Fuel cells operate cleaner because they use electrochemical means to provide power versus combustion. If we implement carbon taxes or carbon caps—which appears likely within the next five years—coal gasification technologies and fuel cells will look even more attractive.

Biomass, which is plant material that contains hydrogen, can be gasified to produce hydrogen. This is an example of a carbon neutral process since plants absorb CO_2 in the photosynthesis process and produce CO_2 in the gasification process. To make the process even greener, the CO_2 can be sequestered and reinjected into geological formations as described for coal gasification.

Other advanced technologies exist, including biological methods to produce hydrogen from algae and bacteria. The challenge here is that the algae will produce hydrogen in the absence of oxygen, but not for long, because oxygen is required for the process to propagate. Enzymes have been discovered that act as a biological switch to allow the algae to produce hydrogen in the presence of oxygen. Types of algae with smaller cell gaps have been developed that keep the larger oxygen atoms from recombining with the algae produced hydrogen. But this isn't yet a large scale viable energy option, and likely never will be.

Love it or hate it, nuclear also could play a part in the hydrogen economy. Due to its low fuel costs per unit of energy and large base load operating characteristics, off-peak extra electrical capacity could be used immediately in electrolysis to generate hydrogen. In the future, the next generation (GEN IV) of high-temperature reactors could be used in conjunction with thermal-chemical process to drive hydro-cracking or reforming of natural gas, or electrochemical processes such as high-temperature steam electrolysis (HTSE). These future designs, like the high-temperature gas reactor (HTGR), advanced high-temperature reactor (AHTR), and gas-cooled fast reactor (GFR)[9] promise to be inherently safer.

Smaller, modular "pebble bed" designs, where the nuclear fuel is encapsulated in graphite balls, also promises enhanced safety. Germany and South Africa are developing pebble bed reactor designs. Nuclear power is seeing increased interest in the United States, even from such unlikely sources as environmental groups. Environmentalists who believe that greenhouse gas emissions are the greatest global concern are more willing to consider nuclear power generation as part of a future energy mix. The lack of harmful air emissions, coupled with hydrogen possibilities, certainly makes nuclear an interesting option. As the world moves forward, it will be interesting to see how God's plan unfolds for nuclear energy in terms of both electricity and hydrogen generation.

Hydrogen Safety

There is nothing like bringing up *safety* to dampen a child's excitement . . . or to quench an excited engineer's conversation about hydrogen. However, safety issues needn't dampen our enthusiasm for hydrogen if we consider it carefully. Unfortunately, the first images that many people envision when you mention hydrogen are the Hindenburg and hydrogen bombs.

Let's start with the Hindenburg. Public perceptions of hydrogen safety usually come around to visions of burning zeppelins falling from the sky.

Figure 6.2—Hindenburg Illustration (But Hydrogen wasn't to Blame!)

But just as Mrs. O'Leary's cow was unfairly convicted of starting the Great Chicago Fire, hydrogen wasn't the primary cause of the Hindenburg fire and subsequent crash. The real culprits were likely its covering and unfortunate weather conditions. For starters, the Hindenburg's skin was a flammable painted fabric, which included many ingredients found in modern rocket fuel, such as aluminized cellulose acetate butyrate. Although the hydrogen used inside the airship certainly did burn, photographs help dispel this as the root cause. Hydrogen burns as a nearly invisible blue flame; the witnesses, photos, and film footage of the Hindenburg show the fabric burning with a bright orange flame even at the start of the fire. Most likely, the fire began on the painted fabric skin and then burned through the hydrogen-filled bladders inside the airframe.

Other facts lend credence to this argument. The stormy weather conditions that day in May, 1937 included lightning strikes near the Hindenburg's landing site in Lakehurst, New Jersey. The amount of electrical charge in the atmosphere would have been sufficient to build up a large static charge on the airship skin. Some witnesses reported the Hindenburg had a blue glow when it approached for landing, indicative of a charge build-up. These weather conditions, coupled with the enormous size of the Hindenburg, which spanned the approximate length of three Boeing 747s lined up tip to tail, set the conditions for disaster. The Hindenburg became a huge charged-up doorknob in the sky, just waiting for an electrical path between it and the ground.

The combination of covering (flammable fabric), conditions (stormy weather), and current (electrical path to ground) caused a deadly trifecta of events to transpire simultaneously. The resulting disaster occurred because of fire that started from the outside in, not the inside out. The moral of the story? Never paint your airship with rocket fuel. Former NASA scientist, Addison Bain, provides more details in his report, *Colorless, Non-radiant, Blameless: A Hindenburg Disaster Study*, if you would like to learn more.

And the hydrogen bomb? It's not built around the standard hydrogen isotope called protium (one proton nucleus). It's actually a nuclear fusion reaction that uses a double-heavy hydrogen isotope, tritium (H_3), that contains one proton and two neutrons in the nucleus. It's a vastly different compound than the garden-variety hydrogen we've discussed thus far. So there's no reason to worry about terrorists stealing a trailer truck of hydrogen and making a hydrogen bomb.

In fact, the safety record of hydrogen is actually quite good; in excess of 10 million tons of hydrogen is produced and used each year in the United States aiding in the production of fertilizer, glass, and ironically, in oil refining.[10] Hydrogen is also used in large electrical power plants for generator cooling. As with any flammable or explosive material, hydrogen should be handled in accordance with safety standards, no more or less than other more common flammable or explosive substances, such as gasoline, jet fuel, natural gas, or propane.

The Burning Bush

> There the Angel of the Lord appeared to him in flames of fire from within a bush. Moses saw that though the bush was on fire it did not burn up. So Moses thought, "I will go over and see this strange sight—why the bush does not burn up."
>
> —Exodus 3:2–3

God can burn something without consuming it, but when we burn a log or a piece of coal, it is consumed and only ash remains. When we combust a fuel like wood to create heat and release energy, we do so at a cost. That's the inherent problem with energy derived from combustible fuels: once that fuel has been consumed, it can't be used again. However, there are alternative methods to deriving energy without combustion, including fuel cells, and I can't help but draw parallels between fuel cell technology and the story of the burning bush. Both show the effects of energy use without combustion. Both show that alternatives to the norm are possible, with God's help. And both show that we, as humans, are curious of the unusual, and sometimes that's how God gets our attention. One thing is for certain: I would like to see the world released from the grip of Middle East oil through fuel cell technology, just as God worked through Moses to release the Israelites from the Egyptians.

Engine Comparisons

Let's take a closer look at fuel cell technologies and how they compare to the combustion engine. For starters, a fuel cell, like the gasoline-powered engine, will run until the fuel is exhausted. Extended run times are available as long as fuel is provided. Due to high pressure, it will presumable take longer to fill the tank with hydrogen than with gasoline, but filling times in the range of five minutes have already been demonstrated.[11]

As with the gasoline-powered engine, there is exhaust. But unlike the pollution from internal combustion engines, the exhaust from a fuel cell is water or steam. Depending on the type of fuel cell, other gases are exhausted, but usually in much reduced volumes compared to combustion engines. The water expelled is very pure; therefore no polluting effects are envisioned for fuel cells. Plus, the steam from fuel cells operating at higher temperature can be used for other purposes, such as heating or absorption chilling (which can be used for air-conditioning).

As they mature, one of the big benefits expected from fuel cell systems is added reliability and reduced maintenance because they have fewer moving parts to mechanically wear out in comparison to engines. But the biggest difference between gasoline-powered engines and current fuel cell technologies is this: combustion engines and turbines are a mature technology, fine-tuned over the last century. It's cheap and proven, but there are no big improvements for the gasoline-powered engine on the horizon. What you see is what you get. Hydrogen, on the other hand, offers a lot more "upside."

Eau de Gasoline

All this may lead you to ask a simple question: If hydrogen is so great, why do we use gasoline today? Good question. Let's pretend that we didn't have the gasoline-powered engine, and that some wise engineer just invented it and tried to market it today. It would fail miserably. Why?

- Gasoline formulas originally contained about 30% benzene, a known carcinogen.[12]
- Spilled gasoline is environmentally destructive to plants, animals and ground water (especially if it contains MTBE).
- Cleanup of spills is dangerous to emergency crews and to the public since it pools and continually emits flammable and explosive vapors.

- It increases greenhouse gas emissions in its production and use, and contributes heavily to smog.
- We import over half of the crude oil used to produce gasoline over long distances from geopolitically unstable regions at the detriment to national security (not to mention the risk of tanker spills and supply disruptions).
- Gasoline is prone to wide price swings due to supply disruptions such as natural disasters, political turmoil, and terrorist attacks.
- The refining process is dangerous and environmentally unfriendly.
- It smells bad. You won't find "Eau de Gasoline" at the perfume counter of your local boutique.

Simply put, lawyers would aggressively sue anybody who introduced gasoline as a new product. Gasoline as we know it wouldn't exist. So why is it so common?

- It is an established commodity.
- The supply infrastructure, "well-to-wheels," has been in place for decades.
- It has good energy density and is a liquid at room temperature and pressure.
- Since it is a mature incumbent technology, it is relatively economical.

The bottom line is that gasoline still offers the best price and convenience. These are good reasons from a purely economic standpoint, but not from a public safety perspective. Being safety-conscious, and also not wanting to contribute to melting his North Pole home with greenhouse gases, probably explains why Santa Claus uses flying reindeer and not an engine-driven sled. I heard he's considering a hydrogen-fuel-cell-driven sled, and letting the reindeer retire.

No Lingering Problem Here

Unlike the Christmas cookies and eggnog that linger around my waistline at Christmas, hydrogen does not linger if it gets loose. The most beneficial safety attribute of hydrogen is its diffusion rate, defined as the speed at which a gas dissipates at normal atmospheric conditions.

Hydrogen has a diffusion rate of 45 miles per hour (20 meters/second), twice as fast as its nearest counterpart, helium. The faster a gas diffuses in air, the quicker it can dilute to harmless concentrations. In an auto accident, this could be a huge safety benefit. Gasoline and liquid fuels puddle up and present lingering problems, as well as propagate a larger fire zone.

For indoor conditions, proper ventilation presents more of an issue, but natural gas has been used in homes for years. If properly handled, hydrogen does not differ much from natural gas regarding its use in public buildings and residences. The safety benefits of a super-fast diffusion rate are a strong plus for hydrogen. In addition, hydrogen causes no pollution to the atmosphere if it escapes, and is not a greenhouse gas. Nevertheless, high diffusion rates aren't always good. Cash leaving my wallet is an example of a high diffusion rate that I don't want to have.

Leak & Spill Effects

If Santa spills the milk left out for him Christmas Eve, he doesn't don a hazmat suit to clean it up. Fossil fuel spills constitute a bigger mess. I witnessed this firsthand in two separate incidents that happened within thirty miles of my house. This is by no means an unusual rate of occurrence.

The first was a number-2 fuel oil diesel leak that ran into a stream. The cost of the cleanup ran into the hundreds of thousands of dollars, but that cost doesn't reflect the intangible costs to the environment and to the local residents who had to deal with months-long reclamation activities just to get back to normal. In fact, it took months, and tons of soil and rock had to be removed.

The second incident was a major gasoline distribution line that spilled hundreds of gallons into the Missouri River before it was shut off. Several square blocks of commercial and residential territory were evacuated. It certainly wasn't beneficial for the environment, either. Similar situations play out through the world and are more common in developing countries, where environmental laws and cleanup are nowhere near as stringent as in the developed world.

Other affiliated gasoline environmental problems include the MTBE additive being phased out of use because leaky underground gasoline tanks have contaminated the ground water. The MTBE additive dissolves in water and poisons the ground water. MTBE does its job well, boosting the oxygen content of gasoline to prevent engine knocking and fulfilling clean air requirements, as long as it doesn't leak. However, leaks inevitably happen and have resulted in millions of dollars squandered in lobbying, litigation, and liability payments since the 1990s, all to protect MTBE from groundwater cleanup costs. This single issue of MTBE litigation protection included in the fall 2004 National Energy Bill effectively killed the bill. Thirty years of overdue national energy policy had to wait another year until the Energy Policy Act 2005 passed, this time without MTBE litigation protection provisions.

On the other end of the spectrum, hydrogen boasts exemplary environmental cleanliness. Hydrogen is not a greenhouse gas, so it does not cause global warming. It does not contribute to smog, acid rain, or any other air pollution if leaked to the atmosphere. Non-toxic and non-poisonous, it will not contaminate land, lakes, or streams. Liquid hydrogen will not either, and quickly vaporizes to a gas at atmospheric temperatures and pressures. Some claim hydrogen is a poor transportation fuel because of the carbon emission problems of fossil-fuel-derived hydrogen, which constitutes 95% of present-day hydrogen production. But in response, central hydrogen producing plants with viable carbon sequestering is much preferred to the carbon that's burned and released from thousands of individual vehicles, where sequestering is practically impossible.

What about natural gas? Although it contains the least carbon and is the cleanest burning of the fossil fuels, it doesn't come close to hydrogen in environmental cleanliness. Natural gas leaking into the atmosphere causes global warming twenty times greater than carbon dioxide. This is due to the heat-trapping effect of the reflective solar energy in the natural gas molecules. A surprising amount of methane is still flared in the process of oil production and in landfills at the rate of 9.2 to 12.5 billion cubic feet per day throughout the world causing additional carbon dioxide impacts at drilling and landfill sites.[13]

Is Everything Heavenly about Hydrogen?

The picture painted so far looks pretty positive for hydrogen. Once it's developed, the requests for hydrogen-powered fuel cells will swamp

Santa's list (and Wal-Mart will find it impossible to keep fuel cell products in stock). Granted, this will take some time, but if people become educated on the benefits of hydrogen and fuel cells, it will happen quickly.

Having stated hydrogen's benefits, even diehard proponents of hydrogen fuel cell technology will admit that it's not perfect.* Hydrogen is flammable in a wide range of mixtures in air, from 4% to 74% concentration. That range is wider than almost all gases, and is most worrisome at the low-end range, because it doesn't take much to get to 4%. However, its most easily ignitable mixture is considerably higher at 29% (stoichiometric ratio), which is much higher than gasoline's 2% ratio, and therefore superior. Hydrogen is also explosive in a wider range than most gases, from 18% to 60% mixture in air.[14] Fortunately, it's very difficult to produce that kind of explosion in high-pressure hydrogen tanks, even if shot with a high-powered rifle, due to the tank's inherent strength and lack of right oxygen mixture.

The other public danger, mentioned in the Hindenburg example, is that hydrogen burns with an almost clear, light-blue flame that is very difficult to see in daylight. And since hydrogen fires don't produce as much radiant heat, it would be easy to accidentally touch off a hydrogen fire. Sensors, odorants, and colorants are being developed to mitigate the hydrogen fire-warning problem.

The other dangers related to hydrogen are high-pressure storage and asphyxiation for gaseous hydrogen. Although challenging, many dedicated scientists and engineers are striving to find workable solutions.

Anything stored under high pressure has inherent danger, and we often store hydrogen at 2,500 to 10,000 pounds per square inch. But because of the high pressure they must contain, the storage cylinders are extremely rugged. Carbon nano-structures, metal hydrides, and chemical hydrides are rushing to provide other options at lower pressures if that proves to be the safer path.

---

* Take love, for example. What could anyone have against love? As with anything, problems come with how love is applied. The Bible teaches the *love* of money is the root of all kinds of evil (1 Tim. 6:9–10). In this teaching, nothing is wrong with love itself, but the love is misapplied to earthly wealth. It is interesting to note that God has no problem with money either, just as long as it isn't loved more than Him and other people. We must learn to focus on the right things, in the right priority.

Any gas in an enclosed area can be dangerous to humans and animals if too much oxygen becomes displaced by the other gas. This includes high concentrations of hydrogen, although the propensity of hydrogen to rise quickly makes the danger of asphyxiation very unlikely. Unless, that is, you are Spiderman, hanging on to the ceiling where hydrogen collects. Cost-effective and reliable sensors are being developed to provide appropriate warning. (Hang in there, Spiderman.) Lesser concentrations of carbon monoxide, a by-product of incomplete combustion inside internal combustion engines and the burning of hydrocarbons, are more harmful because they rob oxygen from the blood stream. Methane released from coal beds is both an asphyxiation danger and an explosive danger for coal miners. Many a canary in the old coal mining days gave its life to asphyxiation to save humans.*

Meanwhile, hydrogen has a relatively safe track record in industrial use for more than seventy years, used primarily today in the production of ammonia-based fertilizers, and enrichment in the refining process of low-grade crude oil. Hydrogen has enormous potential, when viewed against today's combustible engines.

Tag-team Match: Electricity/Batteries and Hydrogen/Fuel Cells

How do fuel cells stack up against batteries? Many believe we should further expand batteries, rather than fuel cells, as a replacement to the gasoline engine. But batteries have inherent problems that will likely never be solved:

- Battery longevity limitations (limited internal supply of stored energy)
- Battery recharge times are too long (and some batteries can't be recharged at all)

---

* Asphyxiation holds a special significance for Christians, since crucifixion causes asphyxiation (not being able to get oxygen into the lungs), and is likely how Christ died ("With a loud cry, Jesus breathed his last" Mark 15:37). Analogous to the canary, Jesus Christ saves our lives too, just in a much more significant way than the coal mine canary. We have to respond and follow the call of the Savior or we will perish also, like unobservant coal miners who ignore the death of the canary and miss the precious chance to be saved. Unlike the canary though, Jesus is risen!

- Battery weight and size penalties (especially a problem for transportation sector electric cars)
- Battery disposal problems (toxic chemicals inside the battery)
- Battery temperature constraints and excessive maintenance requirements
- Upper-end practical size limitations (stationary power)
- Depth of discharge concerns for certain types (battery damage due to energy being extracted too quickly or for too long).

Fortunately, the two technologies aren't mutually exclusive: batteries actually make a very good complement to fuel cells. In fact, the two technologies will likely be used in tandem, especially in the early commercial development stages of fuel cell systems. We've stated already that the primary weakness of electricity is its lack of energy storage. The electrochemical form of electrical energy storage available through batteries is limited. And then there is the inconvenient battery recharge time. As an *energy currency*, it's like regular monetary currency, only good in certain countries.

Hydrogen possesses shortcomings, too. It's presently more expensive to separate from its host molecule, making it an expensive energy currency in comparison to electricity, in most cases. It's also a difficult energy currency to store, per unit volume. Unlike the electrical circuits in computers, hydrogen cannot be used to store information. Electricity enables computer technology. You need electricity to run computers and electronics and always will. Fuel cells will not have the instantaneous response that batteries provide.

But when you marry the technologies—batteries and fuel cells—you overcome the obstacles. What electricity lacks in storage capability, hydrogen can provide. What fuel cells lack in convenience of instantaneous electrical power, batteries provide. Batteries and fuel cells both provide DC electricity, which makes them work in perfect tandem. The next generation hybrid car to hit the market may be a hydrogen fuel cell–battery hybrid. The batteries can provide fuel cell start up support and can help smooth out transient conditions, since battery response times are faster than present day fuel cell technology allows.

Hydrogen can be a material fuel feedstock and stored in huge quantities, which can then be transformed through fuel cells to electricity. Electricity can be used to run motors, electronic devices, computers, and things important for the storage and processing of information. Another area where electricity does well is transporting energy without transporting material. Maybe this does not make sense to the non-engineering minded, but in practicality, the power grid, the existing infrastructure to transmit electrical power, provides a benefit that can be an advantage to hydrogen also. They can work together because H_2 can be converted to electricity, and electricity converted to H_2, they are interconvertible. Both can also be made from any energy source, including renewable sources.

So in the short term, Santa will continue to bring batteries at Christmas, and don't sell your stock in Duracell just yet.

Picking a Winner

Plasma or LCD? Picking the best type of fuel cell for Santa to bring is like trying pick the right television. There are a lot of choices. Become informed and look at the pros and cons of each. To get a more detailed understanding of different types of fuel cells, see the end of this chapter, but here's a quick peek at four of the most popular options:

- Polymer Electrolyte Membrane (PEMFC) fuel cell: Lower operating temperature, faster acting fuel-cell-type that requires very pure hydrogen; targeted for transportation applications, backup power, and small stationary power.
- Solid Oxide (SOFC) fuel cell: Higher temperature fuel cell, targeted mainly for medium to small stationary power; it has a solid electrolyte and, because of its higher operating temperatures, it can reform hydrocarbon fuels internally and use waste heat to improve overall efficiency.
- Molten Carbonate (MCFC) fuel cell: Higher operating temperature fuel cell (but not as high as SOFC); the electrolyte is a corrosive molten material at operating temperatures. It performs similar functions to a SOFC.

- Phosphoric Acid (PAFC) fuel cell: Medium operating temperature fuel cell used in stationary power applications; it has the most operating units in the field, but lately research pace has dropped off for this technology.

On one hand, the high-temperature fuel cell types (SOFC & MCFC) have longer warm-up and cool-down times that reduce thermal stress to the fuel cell components, require more safety precautions (because of the heat), and have heat dissipation inefficiencies if not used in combined heat and power (CHP) scenarios. But SOFCs and MCFCs are more suitable for continuous running stationary applications and where fuel flexibility is important.

On the other hand, PEMFCs are more suitable for small- to medium-size applications like the transportation sector, where fast start-up times and lower operating temperatures are a benefit. In this case, you might not care about using waste heat to improve efficiency.

Each of the technologies above has its strengths and weaknesses. That is why one of these listed fuel cell technologies has not taken off and dominated the others. Since each brings some unique benefit they may all continue to develop.

North Pole Hydrogen Economy (Almost)

Iceland is said to be one of the most unique places in the world. The northern part of the continent kisses the Arctic Circle at the convergence of two tectonic plates in the remote North Atlantic. As a result, earthquakes and volcanic activity are common. Because of the far north latitude, glaciers also form, creating an ice-carved landscape that contrasts beautifully with the volcanic landscape. This results in a land with glacier-carved fjords, moon-like lava fields, geysers, and waterfalls, all of which contribute to Iceland's nickname, "the land of fire and ice."

This breathtaking land may soon have another nickname, "The land of hydrogen and fuel cells," since it will soon become the world's first hydrogen-powered economy. Iceland, as you might surmise, is a challenging place to live. Its thin soil layer doesn't support much agriculture; only 21% is considered livable, arable land.[15] Iceland has no fossil fuel resources, which means no coal, no oil, and no natural gas to use for transportation fuels or electricity. The lack of local fossil fuel is problematic for land, air, and water transportation. It greatly affects fuel for engines that propel Icelandic fishing boat fleets. The fishing industry makes up roughly 20% of Iceland's economy and provides 70% of the country's exports.[16] The cost of engine fuel is a major drain to the local economy.

But how about generating power and heat for homes and businesses? Thanks to Iceland's location on the geological fault line, the internal heat of the earth is readily available. This internal heat is called geothermal energy and provides the "fire" portion of Iceland's nickname. They use geothermal energy to generate 87% of heating needs, and a growing percentage of electricity needs.

The "ice" portion of Iceland's nickname also helps with power generation. Melting snow and ice provide ample water flow for hydroelectric power stations. Nevertheless, the electricity resulting from Iceland's "fire and ice" cannot be used as a transportation fuel for cars, trucks, airplanes, and fishing fleets.

Luckily, the resourceful Icelanders have made overcoming adversity their specialty. That's why they're rapidly moving to institute a hydrogen-powered economy. Given the high cost of fossil fuel imports, the economic benefit of generating hydrogen locally from abundant renewable electricity is better than many other places. The tremendous geothermal, hydro, and wind resources that Iceland possesses to generate electricity can also be used for electrolysis. The hydrogen is captured and can then be used in modified combustion engines, or better yet, in fuel cells. Since hydro, geothermal, and wind are all considered renewable resources, this produces very clean, very green hydrogen, which means that greenhouse gases (primarily carbon dioxide) are not produced. Iceland already has one hydrogen fueling station and several hydrogen-fuel-cell-powered buses as one of the ten cities in the CUTE bus program. I do think they are "cute," but CUTE stands for Clean Urban Transport for Europe. Iceland has declared all city buses will be hydrogen powered within a decade.[17]

Iceland has a perfectly-sized economy for converting to an alternative energy system. Physically, Iceland is about the size of Kentucky, with a population of 300,000 people;[18] 61% of the population lives in the capital city metro area of Reykjavik. All of this helps Iceland, because it's much more difficult to turn a heavily populated country with a large land mass and entrenched energy infrastructure, like the United States, over to another form of energy. Iceland also benefits from having a highly educated population which understands the need for energy change and has the technical wisdom to succeed in implementing it.

This combination of low fossil fuel resources, remote location requiring self-sufficiency, and small but technologically-advanced population make Iceland a perfect test for a fledgling hydrogen economy. Maybe it could be a tempting location for St. Nick's summer home.

Wrapping It Up With a Bow

I've had some fun writing this chapter and I hope you've had fun reading it. One thing I hope you take away from this chapter is this: Santa Claus and hydrogen fuel cells share, or will soon share, a common path of simple beginnings, but with world-changing results. Nicholas of Myra did not set out to become a timeless legend. He simply wanted to make a difference with what he had. Looking to the future, the legend of hydrogen has just begun to change the world, although few today recognize it. St. Nicholas performed acts of charity and kindness to honor God, pure and simple. The genuine spirit of giving with anonymity, expecting nothing in return, generated the notoriety Nicholas attempted to avoid. Long after his death, he is not only part of Christian lore, but also beloved by many around the globe. St. Nick's spirit of giving and love is the essence of goodness, representative of God's gift, Jesus.

Although not comparable to God's gift of Jesus, hydrogen and fuel cells are a gift, too. When partnered with electricity, hydrogen fuel cells can power everything from microelectronics, to vehicles, to cities, while doing it with environmental cleanliness and energy security for everyone, not just those countries with fossil fuel resources. Like the birth of baby Jesus, the intention of this gift is not just for Christians, but also for the world.

The question might come down to this: "What do you hope to find in your stocking Christmas morning . . . hydrogen/fuel cells or coal?" The tradition of St. Nicholas leaving a lump of coal in the stocking of children who have been bad supposedly started in Italy. The threat of coal has since struck fear in the hearts of children across the world. If we're bad, Santa will leave a lump of coal in our stockings, but at least we know we can gasify it and make hydrogen.

CHAPTER 7

DOOM AND GLOOM, OR HOPE AND OPPORTUNITY?

You, my children, were called to be free. But do not use your freedom to indulge the sinful nature; rather, serve one another in love.

—Galatians 5:13

In the previous chapter, we discovered a technology that—with perseverance, opportunity, and hope—can change the world. But that is no guarantee that it will happen. We have been given free will by God, and our country gives us freedom seldom experienced in the history of civilization. What will we do with that freedom, especially when times get tough?

The year 2008 was one of those tough years for most. The nation struggled with financial crises, mortgage foreclosures for many families, and big disparities in food and fuel prices. However, counterintuitive to expectations, the economy experienced record oil price drops in a matter of months as the economy tanked. Oil dropped from $147 a barrel in July to under $50 a barrel during the economic collapse in November. The national average price of gasoline plummeted from $4.26 per gallon in July to under $2.00 a gallon in November. However, the economy went from bad to worse as fuel prices dropped. Therefore, high fuel prices are not a universal evil; sometimes high prices reflect a growing economy and help drive investment in alternatives. The economic meltdown created a river of discontent, despair, and depression, but also created a great opportunity to change course and make improvements. Our government and the economy are going through changes more drastic than at any time in recent history. Avoiding apathy and making difficult decisions will give us the chance to turn defeat into victory.

This scenario has played out many times before in history, when victory seemed unobtainable and people struggled against unbelievable odds. Luckily, the Bible teaches us how to lead our whole life, not just our eternal future in heaven, but also our difficult daily existence here on earth. We should, therefore, derive comfort from Scriptures, knowing that God often works through unlikely means, like Gideon's story from the Scriptures.

The story of Gideon's triumph over the ancient Midianites is a popular offering in Sunday school, found in the sixth chapter of Judges. God called Gideon to lead the Israelites against the oppressors. At first, Gideon balked. "Lord," he asked, "how can I save Israel? My clan is the weakest in Manasseh, and I am the least in my family" (Judg. 6:14–15). But eventually, the warrior Gideon formed a mighty army of 32,000 men. God, however, had different plans and through a series of tests, he winnowed the troops to 300 men, telling each to report for duty with only trumpets and hidden torches. When they encircled the enemy, God told them to blow the trumpets and expose their torches, which so confused the enemy that the Midianites began to fight among themselves, destroying their forces. The rout was complete; all part of God's master plan.

We can learn a lot from this story as we look at our current energy situation, which seems as difficult to conquer as the Midianite army. Gideon succeeded because he concentrated on three key elements: direction, focus, and the right application. Basically, Gideon had a pair of "biblical binoculars."

Biblical Binoculars

I've been told that the Bible points to the cross of Jesus Christ. The Old Testament points forward to the cross; the New Testament points back to it. When you think about it, the view of the cross is like peering through a pair of "biblical binoculars." The left side is the Old Testament, and the New Testament is the right side. Our job is three-fold: pointing the binoculars in the *direction* of the cross; ensuring that we have a proper *focus*, without obstructions; and looking through the binoculars the proper way, using them in the *proper application*.

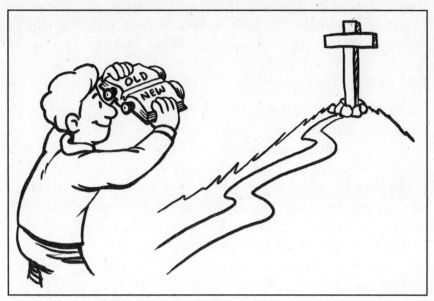

Figure 7.1—Biblical Binoculars Illustration.

DIRECTION

The Bible provides us with a moral compass, always pointing to the cross. From time to time, people will try to point the Bible in another direction and lose sight of the cross. Anyone who has used binoculars knows that once you've lost the image you're viewing, it's hard to find it again. That's why it's so important for Christians to keep sight of the cross; it's also why we stumble spiritually when we do lose sight of it. Gideon succeeded because he took his direction from God alone and looked only to Him.

FOCUS

Anything that blocks our vision of the cross is an obstruction. Sometimes these obstructions are beyond our control, but often sin looms like a huge elephant that we let get in the way.

If we don't have sharp focus on the cross, we will not be able to see the cross clearly. Proper interpretation of Scripture that aligns perfectly with God's will represents sharp focus; misaligned or "fuzzy Christianity" (of which we'll discuss examples) poses a different danger. Gideon didn't concern himself with what his men thought of the plans, or even of the intellectual fear of the disadvantages of fighting a battle with only 300 men. He kept his focus on God alone.

RIGHT APPLICATION

As a kid, it was fascinating to peer through a pair of binoculars in reverse to make things look farther away, rather than closer. But in this case, we want the cross of God's salvation closer to us, not farther away. We often feel the most despair when we feel far apart from God. Keeping our binoculars facing the right way makes sure we feel God's presence and keeps the Scripture message close at heart, bringing us closer to the cross and to God. Gideon defeated the Midianites because he used the tools God gave him properly. Nothing more, nothing less.

As we look at the world today and confront the issues before us, we must ask ourselves: "What purpose does God have for Christians when it comes to energy and the environment?" It is inconvenient, at best, to demonstrate good stewardship these days given the economic situation and the energy-related challenges we confront. To compound our problems, Christians have not united to lead environmentally with the hope and good stewardship message given to us in the Bible. The story of Gideon starts with a common reaction people give when called upon to lead: the usual "Who, me?" response.

But let's take a closer look at environmental, or "creation care" directives in the Bible, which, until recently, wasn't a topic that received much attention or focus. With the Old Testament lens of our biblical binoculars, it doesn't take long to find creation care in the sights.

We find the first examples of proper *direction* of creation care early in the Bible, in the first chapter of Genesis, verse 28: "Be fruitful and increase in number; fill the earth and subdue it." It is the very first command the Lord God gave to man, coming before the command not to eat from the Tree of Knowledge, before Adam and Eve fell into sin. At that point in history, man was in perfect harmony with nature and

God; therefore, good stewardship was at the heart of God's first command to mankind. To completely understand the passage from Gen 1:28, the word "subdue" must be defined and put in context. In some versions of the Bible, the word "govern" or "care for" is used. Subdue can mean "conquer" or "control," but in this case the connotation does not mean "harsh rule," or "forcibly destroy."

Remember the timing of this command: it happened before the advent of sin, when everything was perfect and man was in harmony with God and His creation. God's nature is not to destroy creation, but to nurture and mold it. Sin tries to destroy what God created. The destructive tendency displayed by humans is not a heavenly trait, but Satan's calling card. We should never construe "subdue" to mean that God gave us the earth to abuse and destroy.

Although we are the crown of God's creation, we should not become haughty with that power. We are worth more than many sparrows (Matt. 10:29–31), but God values all of His creation. When God made the covenant with Noah, placed a rainbow in the sky, and vowed never to destroy the world again by flood, God made it also with all life on earth (Gen. 9:8–17). God's creation glorifies Him. It is not our right to senselessly or methodically destroy it for our gain. Instead, our actions should cultivate, care, and create a loving respect for His creation.

Critics say that Christianity lost focus of this fact, and as a result, has greatly contributed to ecological crisis. One of the first to suggest this was Dr. Lynn White Jr. in an article written over thirty years ago titled, *The Historical Roots of Our Ecologic Crisis*. Boiled down, the claim argued that Christian doctrine was complacent in the abuses that Western science and technology rendered and continues to render, and that Christian doctrine trivialized the earth as a resource for our human consumption. The basis given cites man's biblical call of dominion over the earth and the value placed on human life over the rest of creation. Because of Christian theology, Dr. White claims, ". . . we shall continue to have a worsening ecologic crisis until we reject the Christian axiom that nature has no reason for existence save to serve man."[1]

What Dr. White's paper called attention to was a "fuzzy" cross problem. As stated above, our Christian call is not to abuse the earth, but use it wisely and reverently. Although humans are the pinnacle of God's creation on earth, it is false Christian doctrine to state that we needn't care about God's creation, that we are somehow absolved of good stewardship. Christianity's primary focus on our relationship with God in our eternal home in heaven, not our temporary home on earth, may have started this attitude. Of course, our heavenly future is important, but that does not give us the right to misuse the earth and the environment while we're here.

Think of it this way: God has given us the keys to the planet. God created the earth, it is His, but like a teenager who has been given the keys to Dad's car, we should be responsible and respectful, good stewards of the earth that God has entrusted to our care. In being good stewards, we not only show love and respect for God, but also care and love for our fellow man and all of God's creatures.

The New Testament Scripture agrees. The central Christian belief maintains that we are justified (made right with God) through faith in Jesus Christ.* All creation—man, plant, and animal—is made right with God, made possible by Christ's death on the cross and resurrection. Scripture claims this in Colossians 1:20: ". . . and through him to reconcile to himself all things, whether things on earth or things in heaven, by making peace through his blood shed on the cross." It is a gift of grace from God, in no way earned by us. Each individual must choose if he or she will receive the gift of eternal life provided through Jesus Christ. We do not earn the grace of God, but because we have been saved by receiving Jesus as Lord and Savior, we can overcome the sinful world. The rest of creation doesn't have this option to choose to follow Christ, but all of creation awaits with us the renewal at the end of times (Rom. 8:22–25).

What does this mean for our efforts to care for God's creation? It means that we joyfully do the good work of caring for the entire earth God has given us, because Christ saves us, not because we need to earn our way to heaven by caring for creation or any other good work (Eph. 2:8–10). Although sin is still with us on this earth, and although the earth suffers the curse and consequences of our human sin (Rom. 8:18), we still work to honor and glorify God by caring for His creation.

---

* To find out more about how we are saved by the death and resurrection of Jesus Christ, see the following website www.lcms.org or contact a local Christian church to learn more.

We do this not to justify ourselves before God, but because God has justified us through Christ. The earth is a blessing given to us by God, and God has never absolved us from caring for it. The difference is that Christians care for God's creation to show love and respect to God, and in doing so, they show care and love for others. Others care for the environment for different reasons; they do it for the environment's benefit, the benefit of others, and their own self-benefit, without acknowledging God as the Creator of it all.

The Creator God considers it a sin to keep another from knowing Him. God tells us that He reveals Himself through His creation (Rom. 1:20). So when we destroy His creation, we are destroying God's revelation to others, hindering their opportunity to know and see the One who created this beautiful earth. Ignoring God's command to care for the earth not only disrespects the God we should revere, but it also distorts the face of God to others. Granted, we cannot destroy God or His image any more than we could extinguish the sun with a garden hose, but the real damage is keeping others from seeing and actualizing God revealed in His creation, a sin of great consequence.

If the students of a great architect defamed and desecrated his buildings, what would other people think of the students and the great architect himself? They would certainly doubt the greatness of the architect if the students were so disrespectful of their master's work. We commit the same sin when we defame and desecrate God's creation.

This sin not only affects this generation, but also future generations who will see wasteland rather than God's majesty, if we are not careful. Dealing with future problems created by our generation will distract from God's goodness. If people of future generations are battling droughts, floods, heat waves, rising sea levels, and famine brought about by our misdeeds, their focus will likely be more on their basic needs than their eternal soul and relationship with God. Since God gave us dominion over the earth, He will let us destroy it within any boundaries He desires. Why should we test His goodness and mercy? We are better off by far to follow His command, given in Genesis 1:28, which He never rescinded.

An Inconvenient Purpose

Since God has called us to make a difference for Him, we have a purpose. It is that simple, and that complicated. It is personally inconvenient to improve the world's environment and energy situation, but I'm certain it is a purpose God has placed before you and me. In addition, I'm certain the same inconvenient purpose has been placed in the paths of billions of Christians and non-Christians alike. Since very few of the billions burdened with this "inconvenient purpose" are environmental or energy experts, we feel helpless.

The typical answer to the call to lead is, "Who, me?" We are in good company. It's impossible, outside of Jesus, to say that one person in the Bible was qualified to lead or do what the Lord requested. Gideon not only said, "Who, me?" but others—from Abraham, to Moses, to David, to Mary, to Peter, to Paul—have asked the same question. As quoted in the book, *Lead Like Jesus,* Henry Blackaby said, "The reality is that the Lord never calls the qualified; He qualifies the called."[2]

Because we have been called to act as good stewards doesn't mean God needs our works to fulfill His plan. God is more than capable of accomplishing that on His own, but He chooses to work through us and expects good stewardship from us. Failing to answer the call will draw His rebuke. Jesus repeatedly performed miracles *after* an action He requested. He turned the water into wine *after* He asked the servants to first fill the jars with water (John 2:1–11). Jesus fed the 5000 *after* He requested His disciples to bring the five loaves of bread and two fish to Him (Matt. 14:16–20).

It is an honor to serve God's purpose. However, we still often feel overburdened and burnt-out, needing the message of hope. People working on the secular side of ecological and energy matters often have no place to turn with their frustrations, whereas the Christian has the joyful benefit of being able to turn his troubles over to God. Not that God will remove all the obstacles entirely, but He will lighten the burden on our heart. The inconvenient purpose of good stewardship will be lightened by God's grace and the pursuing of alternative technologies.

Uncle Sam to the Rescue?

Just as we can't solve the world's problems on our own, I don't believe we can trust national governments to solve the problems for us. This isn't an issue of republican vs. democrat. Energy and environmental issues transcend political parties, borderlines of countries, and short-sighted economic planning. Energy and environment affect everyone, and they are very much a "God thing."

How can we hope that government will do the right thing? Will their actions be timely enough to help? How can we trust government to get it right when they're often so inefficient? Many feel compelled to turn to help from the government. Whether this stems from the feeling that elected officials should solve problems of society, including energy-related problems, or from a belief that the government is better equipped to deal with large-scale issues than individuals, many agree the government should be involved. But the question remains, at what level? Luckily, our hope doesn't rely solely on government, or any of the other usual places we seek answers, such as Wall Street or environmental activism.

In the United States, reaching an agreement on the proper level of government involvement means hotly debated, continual tug-of-wars between democrats and republicans. Eventually, they reach compromises that both groups can live with, although neither is completely satisfied. Then the next bill and the next congress comes along and they do it all over again. To quote Newt Gingrich on climate-change wrangling, "What you have is the people on the right know they're against regulation and they're against taxation and they're against bigger government. They don't want to think about it because the only answers they receive are things they hate. People on the left know the environment's important, but their answers are all regulation, taxation, litigation. And so we're caught in this gridlock because the left insists on pain and the right insists on avoidance."[3] How true. Is there hope for the political stalemate to change?

Don't get me wrong, I'm not all doom and gloom about American politics. What grieves me is the lack of practical political discourse on both sides of environmental and energy issues, such as the climate change debate and Middle East conflicts, to name a few. A large part of the reason national energy policy is only recently receiving political attention is because it requires tough decisions and a long-term strategy. Neither of those characteristics endears energy policy to politicians. They're usually more concerned with short-term issues like the next election, and answering their constituents' cries of "What have you done for me lately?" In addition, many emergency short-term crises divide attention from long-term issues. Maybe the best we can do as Christians is forgive them for their shortcomings.

At the heart, energy issues, like everything else, belong to God. We should seek answers in Scripture and prayer, but we seem bent on solving the problems on our own or through the government. God, of course, has shown repeatedly that He can work through governments as well as individuals, so politics will likely factor in, but only as a distant second to reliance on God. Help from the government, federally or locally, should be appreciated, but not expected. It will be icing on the alternative energy cake. Frankly, I think many would be happy if the government would stop subsidizing and giving tax breaks to the established energy industry for mature technologies. Since Washington or state government may never construct a good energy policy, what we really need is good energy products, such as fuel cells. America needs to be prepared to do this outside of Washington. If the politicians get their act together along the way, all the better for everyone.

Clear Path Aspects

Another challenge for creation care is not becoming overwhelmed trying to save the planet. That is not our prerogative. Our goal is to leave the saving to Jesus, and serve God by caring for His creation. We risk losing the clear path to the cross by getting too bogged down in the details.

At times we lose the clear path to the cross by allowing other idols to stand in the way, such as preoccupation with fossil fuels and other fuels that harm more than help God's creation and, arguably, humankind. The oil contained beneath those predominantly Islamic countries fuels wars and geopolitical tension, at least in part.

We get distracted from our clear path in many different ways. The amazing story you see over and over in the Bible is how God molds people to his purpose, regardless of their own plans and conventional logic. It's true of Gideon, Moses, young David when he slew Goliath, and the New Testament when Jesus called ordinary fishermen to be His disciples. In each, God knew the virtues required to succeed in the person, even though earthly wisdom would say otherwise. Everyone struggles with God's will, but if one honestly seeks it, God will reveal it. *

We should rightly explore the wonders of discernment of God's will. Proverbs 14:6 states: "The mocker seeks wisdom and finds none, but knowledge comes easily to the discerning."

Discernment is one of the wisest gifts of God we can pray to receive. If we're confident that we're living in God's will, we will be able to find peace even in the difficult times, such as the energy conundrum in which we are involved. Mistakenly, many believe that by living in God's will we will escape those difficult times. That is not the case. Daniel wasn't spared from the lion's den, and even Jesus wasn't spared from the cross. What we can obtain, though, is inner peace that only God can provide.

Marc Gunther, in his book *Faith and Fortune,* discusses the topic of discernment, including how it relates to real-world business decisions:

> "Would God really have an opinion about whether an entrepreneur . . . should sell his business? If so, could a business executive try to discern whether to open a new plant or develop a new product? For that matter, could an investor discern whether to buy or sell a stock? I couldn't see how a religious practice that is inherently rooted in mystery could be of help in making pragmatic business decisions . . . . my initial reaction reflected the conventional wisdom that the sacred should be kept apart from the secular. People who bring their faith to work say instead that the sacred-secular divide is a false one that must be bridged."[4]

The sacred-secular divide falsely blocks our clear path to the cross. Gunther remains objective in *Faith and Fortune,* but the Bible clearly

---

* "Ask and it will be given to you, seek and you will find; knock and the door will be opened to you. For everyone who asks receives; he who seeks finds; and to him who knocks, the door will be opened" (Matt. 7:7–8).

states that God created all things; therefore, why would we want to establish a separation? (Jer. 10:12). This brings the discussion to an important point: since God is truly active in this world, as I believe He is, shouldn't we ponder God's will more often regarding our day-to-day activities? I'm not saying we should pray for discernment of which flavor ice cream God wants us to choose, but important decisions in business, in energy choices, and in environmental decisions should all be taken to God in prayer.

God has dominion over all things in heaven and on earth, correct? Many people struggle with the same issues, carving only a small block of time for God or spiritual stuff in their lives. It's a constant battle not to give in to the secular mindset. *What does my boss think? What will my coworkers think?* We suffer the desire to fit in, the fear of taking an unpopular stance, and the crippling fear of losing a job.

Likewise, for energy decisions, there is a sacred-secular divide. The problem with the sacred-secular divide is so deep, even recognizing that it exists is a challenge. It is very easy to think of energy in only economic terms, like "How can I save on my electricity bill next month?" or "Where can I find the cheapest gas to fill up my car?" There is nothing wrong with asking these questions. Lacking are the less obvious questions, the questions few stop and think to ask, often because the status quo is so familiar and so taken for granted; questions like, "How much am I affecting the environment by the amount of electricity or gasoline I'm using? Where does the gasoline or electricity I'm using ultimately come from? Do I have alternatives? How do I become *God-centered* in my life, including on energy?"

These difficult questions involve some soul searching and a commitment to change. The massive size and extent of energy infrastructure refutes the possibility of overnight change. It's easy to feel overwhelmed and helpless, even if you commit yourself to change. Much like Gideon, who felt small and weak leading a fight against a mighty nation, we can feel insignificant compared to mighty oil companies, giant utility companies, governments, and sometimes even within our own churches and families. Trusting in God, as Gideon eventually trusted, provides us with an example of hope in the midst of hopeless odds.

Another impediment to a clear path is the fear of putting environmental care ahead of people care. But to quote Ed Brown in *Saving God's Green Earth*; "One of the arguments I get into is the position that we need to care for people over the environment and that people are more important than plants and animals. My response is this: We as people, depend on the environment to live. When the land goes bad for whatever reason, human life stops. You can't say people are more important than the land because it's a false dichotomy. When you care for the land, you're caring for people."[5]

As long as we remain focused on worshipping the Creator, and not the creation (Rom. 1:25), we can balance the need to respect the creation while showing love and care for people, too. If we focus on caring for creation, we need look at how to best accomplish this, because there are many ways. Unfortunately, human nature relies too much on the element of fear to accomplish those goals.

Breaking the Cycle of Fear

Those who predict climate catastrophe use fear and incite worry in an effort to motivate and scare people to action. They use "the climate hammer" much like fire-and-brimstone preachers use fear to bring non-believers to Jesus. The intentions might be noble, but the tactics are questionable. A Christian view can inspire action through desire to please God and show love, without the push and pull of worldly fear. Small, but important, steps such as recycling, energy conservation, and energy efficiency can be started immediately. Longer term energy supply actions, such as fuel cells, can be encouraged.

The command given most often in the Bible is not to be afraid. It's given 366 times.[6] Fear is a strange emotion. Our feelings of fear can encompass everything from fear of the largest creature, such as whales, to the smallest, such as bacteria. It ranges from rational to irrational, but regardless, it's very real to the person who has a particular fear. If anything, hysteria that we are destroying creation is misguided. We should only fear the wrath of the Creator, not the creation. But destroying the creation disrespects the Creator. As Christians, we should be wary of this. It is not a political battle. It is not a battle of placing the temporal earth over the eternal human soul. For Christians, it is a choice between following God's command to care for the earth or ignoring it. What the Bible does tell us to fear is not what can kill our body, but what can take our eternal soul.

Personally, though, I believe the largest stumbling block to confronting our energy problem isn't fear, but apathy. America is slowly awakening to the necessity of good energy policy, good stewardship, and personal responsibility regarding environment and energy decisions. However, apathy remains within a large sector of the community. It is encouraging that topics involving environment and energy are receiving more attention, but doubts remain in many minds regarding where truth lies on issues, such as human- or naturally-induced climate change, and how to handle the oil-rich regional turmoil in the Middle East. We have to clearly see the problems in front of us before we can begin to fix them.

Footprints

So now that we know God's plans for us as good stewards (direction), and have removed obstacles such as sin, fear, and apathy (focus), let's turn to proper application. The scales are tipping towards secular environmental activism, which may contain some well-intentioned goals, but is an inferior path. Should we voice concerns over this path and risk derailing environmental protection progress? Is the environment the best gauge to use? The truth is, environmental issues are really energy issues at their core. Energy, and more specifically how we use energy, determines our environmental *footprint*, or the environmental cleanliness scale popularized in the media. Most often, you hear about the *carbon footprint*, but that's really just one metaphorical type of shoe used to visualize the amount of carbon we're using in our lives. Right now most of us are making carbon footprints with clown shoes, not ballet slippers.

More important, however, is our overall *energy footprint*, which should incorporate both the size and depth of our footprint. Currently, our energy footprint is the sum of our energy use, beginning with extracting energy resources from the ground, to refining it, to using it, to the waste products that remain. Many things go into controlling the tangible size of the footprint, including measurable quantities such as carbon, pollution, efficiency, and conservation. The new dimension I'd like to add to the footprint analogy is depth, the intangible impact made by our steps. The deeper the footprint, the worse the impact to our environment, including intangible factors such as how it affects the poor, other species, and future generations. Thinking of it this way, stepping in mud leaves a much deeper and lasting footprint than hard dirt. Intangible benefits include adherence to Christian and moral ideals, increased national security, increased energy independence, reduced conflicts over natural resources, reduced need for troops in the Middle

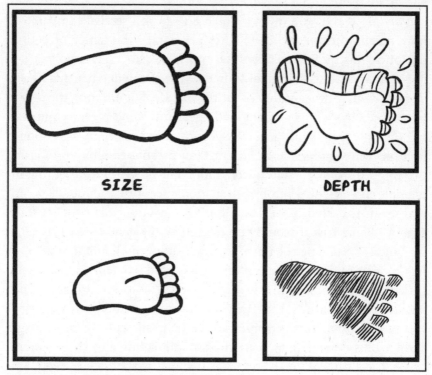

SIZE   DEPTH

Figure 7.2—Carbon Footprints Illustration.

East and military policing, cleaner air and water, reduced health risk, and increased public acceptance.

Conservation means doing without, or reducing our consumption, rationing the available energy resources. Unfortunately, modern society is all about consumption. Too often, waste in the name of personal convenience becomes a measure of wealth or status. The unnecessary excesses, attained easily and comfortably afforded, often mark success. The commercial mindset that more is always better makes it seem un-American to espouse conservation. But holding something back for future generations is a noble cause. And conservation measures not only reduce the measurable size of our footprint, they also have both positive and negative intangible impacts to the depth of our energy footprint. Still it is hard to convince others and feel that self-restriction is an improvement to our lifestyle. Conservation should not be confused with efficiency, which is a different concept in factoring our energy footprint.

Efficiency is the ratio of how much energy you receive compared to how much you put in. The higher percentage the better of course, but often this measure receives too much weight. It gives no value to the cleanliness of the original source or the intangible benefits of one energy resource over another in the big picture. For instance, the overall process efficiency of generating hydrogen from wind power, compressing the hydrogen, storing it, transporting it in a pipeline, and using it in a fuel cell vehicle is less efficient than making gasoline from crude oil supplied from an established Saudi Arabian well. However, the intangible costs to the environment, national security, and morals do not end up in the efficiency calculation. On an intangible benefits scale, with 10 being best, I would rate the renewably generated hydrogen a "9," and the Saudi petroleum a "1." Iraqi supplied oil might be a "0" or less because of what it cost our servicemen and women.

Fuel type determines much of the energy footprint size and depth characteristics. Coal, for instance, creates a larger footprint because it is a more carbon intensive fuel source than oil, causing vastly larger carbon dioxide emissions, particulates, and sulfur dioxide emissions. Oil has less carbon; nevertheless, it is much worse on the depth impact because of geo-political, national security, and moral trappings. Nuclear power produces almost no carbon emissions and requires small amounts of uranium fuel for large amounts of electrical output. However, radioactive waste concerns, along with terrorist and natural disaster possibilities (warranted or not, given nuclear energy's impressive safety record in the United States) will always involve doubt and moral dilemmas as to whether it should be used. Public sentiment in favor of nuclear power, recently shifting in America, will always be just one disaster from disfavor and another twenty years of no-build policy. The size of the footprint is small, but the depth is moderately deep for nuclear power because of the nuclear waste issues. That is why hydrogen provides the best solution. It results in the smallest combined footprint, size and depth.

The bottom line is that God wants us to use His creation that He has given us, but He wants us to share and use it wisely. It's the loving story that Jesus has asked us to live. Our hope lies in Scripture. How can the Bible, which hasn't changed in 2000 years, provide answers to a modern problem that requires technical solutions? The Bible isn't a chemistry textbook, but it does give us the structure to address problems such as climate change, dependence on foreign oil, and national security with amazing clarity. What is more important than scientific breakthroughs the vision the Bible provides, exemplified in Jesus. Jesus said in the Gospel of Mark that the greatest commands are first to love God with all your heart, mind, and soul; and secondly, to love your neighbor as yourself. How can we show love to God by wrecking His creation with our gluttonous appetite for fossil fuels? How can we show love to our neighbor by harming the environment around him and hoarding his natural resources, such as oil?

CHAPTER 8

ROADS TO FREEDOM

"All truth passes through three stages:
First, it is ridiculed;
Second, it is violently opposed; and
Third, it is accepted as self-evident."
—Arthur Schopenhauer[1]

There has long been a love affair with the American road, and the freedom, mystery, romance, discovery, and adventure it inspires. But there is also a practical side. These arteries and veins transport a lifeblood of goods and services around the country. When President Eisenhower made the construction of the interstate highway system a priority of his administration, it wasn't to promote the sale of sexy convertibles, to feel the wind in our hair. The primary driver was civilian requirements, but it also included national security needs. As an officer stateside in the United States Army, Eisenhower saw firsthand the poor conditions of our cross-country roads. After experiencing the German autobahn in World War II, Eisenhower concluded that America needed infrastructure improvements to transport essential supplies, to enhance commerce, and to move or evacuate people quickly in case of a nuclear or other threat. The wind-in-the-hair benefit was a freebie.

Roads are a perfect symbol of the energy picture. Some roads are short and straight, some are long and winding, and some circle back to the same place. Some roads simply end. They are roads to nowhere. The road to energy sustainability and independence is under construction, but in reality, it is not about independence, but about dependence on God. There are many potential roads to energy independence; most are dead-ends or a mirage. The vehicle we take, if we choose wisely, will be powered by dependence on God. Many people might disclaim the need for dependence on God, but denying God's truth never does anything to change God's truth. What road are we on? Does it lead where we want? Are we willing to let go and let God take the wheel?

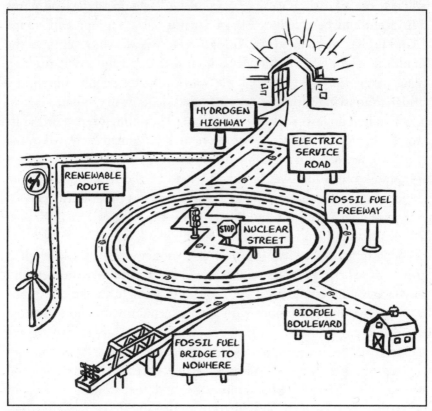

Figure 8.1—Roads to Sustainable Energy Illustration.

Road to Nowhere

In the early 1980s, my parents and I took a trip to Hot Springs, Arkansas. Mom took the wheel while my dad and I napped. Instead of

Hot Springs, we ended up in Crow, Arkansas, one-and-a-half hours off course. That's what can happen when you don't have a map or familiarity with local directions and get no help from your passengers. Likewise, that is what can happen to us if we don't start seriously contemplating our energy future. We end up lost and further away from where we want to be.

Continuing the status quo and calling it an "energy plan" only drives us further off course, propagating the myth that our oil supply is endless, that economic markets supply all the answers, and that the environment will take care of itself. If we're not careful, we'll end up on the road across a bridge to nowhere, like ex-Alaska Senator Stevens' proposed Gravina Island Bridge, nicknamed the "Bridge to Nowhere," which regained the public spotlight when John McCain chose Sarah Palin as his running mate in the 2008 presidential election. To many people, building a bridge to an island with fifty permanent residents at an estimated cost of $398 million dollars exemplified pork barrel government spending at its worst. Granted, differences of opinion are discriminately treated in the media sometimes, but building infrastructure just because we can does not always mean that we should. It has to be the right infrastructure project, benefiting the country as a whole, not just for political gain.

If we're not careful, we could end up like the King of Tyre in Ezekiel 28, prideful and vain in our good fortune and wealth, only to come crashing down to reality. The time and money we spend developing conventional energy infrastructure could be in vain. We could end up buying gasoline at $20 a gallon, frantically trying to make the transition to a hydrogen economy, with a world economy in shambles and a decimated environment. Until more people realize *and care* that grave dangers lurk on this road to nowhere, we'll never be able to see through the fog of apathy and self-absorption that keeps us blinded. All too often, CEOs, stockbrokers, politicians, and the growing throng of Western-style consumers entrench themselves in lifestyles of materialism and power enabled by cheap energy. Trapped by shortsighted gain, people often miss the opportunity to develop the much richer and rewarding long-term gain that is possible with sustainable energy.

Alexander Green, writer and editor of *Spiritual Wealth* says it well: "Money. Possessions. Luxury. These are not the hallmarks of a life well-lived. At best, they are merely by-products."[2] A life well-lived requires more than acquiring as much as we can, including cheap energy; a life well-lived is reflected in what we leave behind and what we create. Paving the way for a sustainable energy future is much more rewarding, and does not preclude the availability of creature comforts. If anything, it will enhance them as a by-product of a life well-lived.

Since sustainable energy is central to economic success and a healthy environment, getting off the road to nowhere is crucial to future generations. We need more than a plan—we need a vision. We must improve our attitude, demonstrate strong will, provide good leadership, and most of all—we must accept divine guidance only possible through God's grace. We need to avoid building a modern-day Tower of Babel (Gen. 11), where humanity unwisely planned to build their way to the heavens. We must respect God's will and His creation in our plans to build the road to a new energy economy. But first, we must explore the roadmap of possibilities that exist.

Fossil Fuel Freeway

The road of fossil fuel is much like the super highways that encircle big cities. They are big, important, and fast, but unless you get off of them, all you do is go around in circles. For us to go anywhere, we have to get off the fossil fuel track. It bears repeating again and again: a mental shift must happen regarding the phase-out of fossil fuels, especially oil. Only then will actual change happen.

Fossil fuels have been the path to immense economic growth for the western world. In all fairness, the upper and middle-class lifestyles many enjoy were built on the back of the environment, yoked to cheap fossil fuels. The unmatched abundance, cheapness, and energy density offered by fossil fuels made them impossible to resist. As with all things this side of heaven, some good, some bad, and some unintended results came from our use of fossil fuel. To our ancestors' defense, the full weight of future environmental and social burdens was almost impossible to predict due to the industrial age population explosion. World population was estimated at the start of the industrial revolution (circa 1780) at approximately 850 million people. Today, estimates exceed

6.7 *billion* people; future estimates expect to reach in excess of 9 billion people by 2050, according to the United Nations. In some respects, the industrial revolution went too well, leaving future generations ill-prepared to juggle simultaneously the population boom and unsustainable fossil fuel use. The industrial revolution wheels are still in motion. Will the tires stop by sudden blowout, will they go flat via slow leak, or will they manage to run until we can change them out for a new set?

Many of us have our tires in a rut when it comes to energy. Expanding our fossil fuel dependence by drilling for more oil and gas and mining for more coal is like stepping on the accelerator, spinning the tires, creating a deeper rut. Drilling in the Arctic National Wildlife Reserve (ANWR) for domestic oil supply only delays the inevitable, putting the country further out on the limb of oil dependence. The recoverable estimates of 16 billion barrels[3] in ANWR would only supply the United States for little more than two years at current demand of 20.7 million barrels a day. When factoring the costs of a new infrastructure to the remote area in northern Alaska, the environmental impact, and propagating further dependence on oil, balanced against the benefit of new domestic supply and potential pump price reduction, the reward is marginal.

Offshore drilling suffers the same setbacks, more or less. Proponents of drilling say it will increase national security by reducing dependence on foreign oil from the Middle East. That is a worthwhile endeavor, but it does nothing to address the larger oil dependence issue. It's like changing supply from a crack house in a bad neighborhood to a crack house in a good neighborhood. The bottom line: the addict is still addicted to crack.

The development money for ANWR at this time would be better spent more aggressively developing fuel cells and a new hydrogen infrastructure. Over 30% of oil consumption in the U.S. factors into products other than transportation use.[4] The need for oil will continue long after we stop using it for transportation. Therefore, we should conserve as much as possible for future generations. Extraction technology will continue to improve environmentally and in efficiency. Then, when we really need oil, we can develop it more safely, more efficiently, and for a smaller basket of products.

However, slamming on the brakes will not get the wheels out of the fossil fuel rut either. Like it or not, fossil fuels provide 70% of the electrical power generation in the U.S. and 96% of the transportation fuel.[5] We cannot afford to quit cold turkey. We need to rock gently out of the fossil fuel rut, changing direction before we can move forward. Since people have grown accustomed to reliable electricity and mobility freedom due to cars, it requires more than technological breakthroughs to move to sustainable alternative energy. It requires a change in mindset. Fossil fuel dependence is the only energy realm the last four generations really know and love. To change the mindset, consumers have to reconsider how to use fossil fuels—to move past fossil fuels.

Fossil fuels have some new roads under construction, namely oil sands, oil shale, liquefied natural gas (LNG), and coal gasification. Although unsustainable as long-term solutions, they provide time to move to other alternatives. What cannot be stressed enough is that these fossil fuel options are not the end, but only the means to an end. These unconventional fossil fuel sources and technologies are one way we can start rocking back and forth to get of the rut we're in, but they should be abandoned as soon as possible when sustainable alternatives such as hydrogen fuel cells are better established.

The most important and extensive oil sands ever found reside in the Athabasca Basin in the lands of our friendly Canadian neighbor to the north. The size is impressive: approximately 173 billion barrels estimated as recoverable by the Alberta provincial government in 2007.[6] They are close by, miles instead of oceans apart from the tradition conventional supplies we currently import—and best yet, they are not in the hands of oppressive Middle East governments that dramatically endanger our national security.

The problem with oil sands is not quantity available, but recovery and delivery of the resources. The amount of energy input required from natural gas to free the oil is astounding. Vast amounts of natural gas are combusted to create steam, which when injected into the earth loosens the thick oil from its geological prison. This process is called in-situ heating. Other serious considerations weigh heavy on oil sands production, including the refining and hydro-treating costs, the copious amounts of water, land, electric power necessary, and the diluting

143

agents required to make the oil tar (bitumen) flow. And it all impacts the ecological landscape.

Oil shale, which is located in the Green River area of Colorado, Utah, and Wyoming, shares similar ecological problems with oil sands. If implemented, in-situ water injection problems would occur, increasing the salinity of ground water and requiring two barrels of water for every barrel of oil produced, which would exacerbate the Colorado River basin water shortage problems.[7]

Liquefied Natural Gas (LNG) is beneficial because it captures waste gas that traditionally has been flared (burned as a waste by-product), and instead makes it useful as a fuel source in easily transportable liquid form. But it takes more than 30% of the energy content of the raw natural gas (methane) to liquefy it, and then it also takes energy to expand it back out to a gaseous form at its destination terminal.[8] Perhaps the biggest problem with LNG is that the largest reserves of natural gas reside in the former Soviet Union and the Middle East—precisely the geo-political areas that we are trying to reduce dependence on in order to increase our national security. In addition, building LNG receiving terminals is challenging because few coastal areas want them built in their backyard.

Coal gasification is a process first developed by the Germans in WWII because they lacked access to petroleum reserves. The Fischer-Tropsch process was used to turn coal into liquid fuel to power the German war machine. South Africa also used gasification as a fuel source when the rest of the world ostracized them during the days of apartheid.

Gasification technology makes a comeback whenever high prices appear and uncertainty of conventional oil and gas markets threaten. The potential it provides for capturing carbon dioxide is an added benefit. The power generating technology called Integrated Gasification Combined Cycle (IGCC) combines capabilities that include capturing (and potentially sequestering) carbon dioxide before combustion instead of afterwards, when it is much harder to separate from other combustion by-products.

It is also an avenue to produce many commercially valuable gases, including hydrogen. The difficulties lie in the expense and extra energy required to perform the gasification process, as opposed to just burning the coal directly. This drives up the amount of coal required to produce an equivalent amount of usable power. The penalty is approximately 20% extra energy required for IGCC, and can reach higher than 40% for traditional pulverized coal plants, if capture of carbon dioxide is included. Storing the massive amounts of carbon dioxide away (carbon sequestration) increases the costs even more and long-term effects are unproven, but the price could be even higher if we continue to release carbon dioxide at present rates.

So what is the solution? Truly, I do not think anybody knows, which is really inconvenient. However, in the short term, capturing and seques-tering at least some of our carbon dioxide emissions to slow the rate of human-caused atmospheric carbon represents the wisest road to take, because it helps take our nation's foot off the fossil fuel accelerator.

The dangerous Siren's song of fossil fuel lies in the fact that it's still abundant, luring us to the rocky shores of energy dependence. Quantities of coal, oil, and gas are plentiful for the time being, but not endless—especially at the rate we use them. The more we build and increase the infrastructure that uses them, the longer we go around in circles around the freeway, failing to build off-ramps to alternative energies. Evidence of dwindling conventional supplies is found in the increase of Natural Gas Liquids (NGL) required to keep the Saudi supplies at their desired production level. This helps mask the fact that crude oil production is actually dropping, indicating that we may be near the peak of conventional oil. If the peak oil theorists are correct, and we are truly moving past Hubbert's Peak (see chapter 2), the price of oil could escalate even more rapidly.

According to EIA statistics, in January 1994, the average price of a gallon of gasoline hovered around $1.00. The first monthly average $2.00 gas occurred May 2004, and the first $3.00 threshold broke through in July of 2006. In February of 2008, gasoline averaged $3.08 per gallon. Less than five months later, gasoline exploded past the $4.00 average mark for the first time, approximately 94 cents differ-ence in less than half a year! In a time frame of less than fifteen years, with inflation relatively in check and without appreciable increases in gas taxes, gasoline went from a $1.00 to $4.00 per gallon. The price is shocking, but perhaps more astounding is the price volatility. In the calendar year 1994, the monthly average price differed no more than 16 cents for the whole year, a 16% change in price. In comparison, 2007 saw a range from $2.29 to $3.19 per gallon, a 39% change in price spanning 90 cents. The volatility demonstrated in the 2008 price spread of 94 cents in less than five months roars in with a 31% change, benchmarked on the first $4.00 weekly average crossing. No wonder the American consumer wants off the roller coaster of gasoline prices.[9]

The answers do not lie with drilling in ANWR, or continental shelf deep-ocean drilling, or oil sands and shale. Even if started today, noticeable results could take five years or more. Considering reserve quantities estimated and the difficulties in extraction, not to mention environmental impacts, answers cannot rely on a crude oil future. The princes in Saudi Arabia would love you to believe oil is the future, just as they would love you to believe that Mohammed is the path to heaven, not Jesus. We need sustainable solutions and we need to develop them in earnest, soon. Hope remains as long as we rely on the Lord and focus on building off-ramps from the fossil fuel freeway, not adding lanes or adding on-ramps to the fossil freeway that only goes in circles, without really going anywhere.

The traffic jam on the fossil fuel freeway cannot be cleared up overnight. It carries heavy traffic and requires a new infrastructure before we can evacuate. But if we act wisely, we can build off-ramps to alternative and diversified energy sources. For example, developing IGCC (especially with carbon sequestration) is one off-ramp. If oil companies would spend more money and time on developing alternative technologies, less time and money on political lobbying for the status quo, and less time and money on developing unsustainable oil infrastructure, then we could create even more off-ramps.

Renewable Route

Renewable energy sources such as solar, wind, wave, tidal, hydro, biomass, and geothermal, to name a few, provide dreamy visions of clean and sustainable electricity generation, both large- and small-scale. Renewable energy also provides hope for transportation via bio-fuels (discussed in a separate section), solar, and indirectly by renewable electricity. But what defines the route to renewable energy?

According to the U.S. government Energy Information Administration (EIA) definition, renewable sources are "Energy resources that are naturally replenishing but flow-limited. They are virtually inexhaustible in duration but limited in the amount of energy that is available per unit of time."[10]

Renewable sources provide hope for energy independence from unsustainable entrenched technologies. It also means much more than that—it means hope that we can change, that we can *renew* our call to stewardship of God's creation. But like the EIA definition, we are limited, also; limited by what we are capable of at this moment, due to our sinful human condition. The Bible states that the earth will not be fully restored until Jesus returns. What's available to us in abundance is God's grace—we cannot renew ourselves, but we can (through the Holy Spirit and prayer) renew our hope for the future. What we need to follow our renewal is *storage*. For renewable energy, we need storage to counter-balance the *flow-limited* restrictions of intermittent power. For our renewal of godly stewardship, we need storage of hope in our hearts, because the daily grind of life makes biblical stewardship a difficult route.

The renewable route remains a difficult road, like a country road far away from the city. Often the renewable route is not well-marked, well-traveled, or well-known. This is much like renewable energy today. It is not well-marked on the roadmap of incumbent technologies; it is not well-traveled, since only a small percentage of our energy comes from it; and it is not well-known, since few people are familiar enough with it to be comfortable of the route. In order to understand the renewable route, we have to understand better the two types of vehicles traveling on it: renewable energy for transportation use, and renewable energy for stationary electrical generation (and future hydrogen generation).

TRANSPORTATION

Renewable energy benefits transportation from the possibility of plug-in hybrids charged by renewable generated electricity, or alternatively, vehicles powered by fuel cells supplied with renewable hydrogen, generated by electrolysis from renewable sources. The other renewable alternative—solar power for vehicular transportation—remains limited by the physical area and size needed to produce adequate power. When solar-powered cars couple technologies with advanced battery technology or fuel cells, the picture brightens. The solar cells do not store energy, which makes energy storage a priority. That's where rechargeable batteries and generating hydrogen for fuel cells make sense. But even with energy storage, there is not enough surface area and solar panel power density to make the leap to solar-powered vehicles. However, solar power can potentially provide auxiliary power for the vehicle, to run its sound systems, navigation systems, and other electrical loads, thereby reducing the battery load and recharge times.

The technology known as solar photovoltaic (PV) holds the greatest promise because of the flexibility in the number of potential products, automotive and beyond. If part of the PV family (known as thin-film solar) proves viable in cost and amount of power generated per square inch, solar power will play a part in automotive and a host of other portable technologies.* The path to the Holy Grail for renewable power lies in the effective and cheap storage of energy generated by renewable sources for transportation and stationary power. If skyscrapers, building roofs, and even homes could incorporate thin-film solar technology, millions of kilowatt-hours of energy from the sun that go unused could be put to work providing sustainable energy and preserving precious resources for future generations.

STATIONARY POWER

For stationary power, the renewable route shows the way to a better future. Wind power produced at comparable cost to the fossil and nuclear workhorses makes it a breath of fresh air and the darling of the renewable energy movement. And the cost of wind power (7 to 10 cents per kilowatt-hour) makes it attractive compared to fossil fuel. If

---

* Thin-film PV even works if fabricated into clothing to provide power to digital music players and such. Your "power suit" can actually be a power suit.

carbon dioxide is taxed, or market-based carbon restrictions (carbon cap and trade systems) come into play, wind will become even more attractive.

But not only will wind energy benefit from subduing carbon emissions; all carbon-emission-free alternatives will benefit, because investment will rush to these greener alternatives. Carbon dioxide has benefited from its omission from environmental protection legislation the whole duration of the industrial age, until recently in 2007, when the Supreme Court ruled that the EPA had the power to regulate carbon dioxide emissions under the provisions of the Clean Air Act, if the government found that greenhouse gases were indeed harming public health and welfare. In April of 2009, a momentous decision was made by the EPA that declared carbon dioxide and other greenhouse gases responsible for the endangerment of public health and therefore under government regulation authority if it chooses. Politicians who refused to act for years on legislation, now face the politically worse threat of regulation outside of their control, and likely will move forward with new laws establishing a cost for carbon. Just because a value hasn't been assigned to carbon emissions in the past does not mean carbon has no impact on the environment. It only demonstrates our lack of past knowledge and the difficulty of assigning a reasonable economic value to indirect environmental effects. Proponents and opponents of carbon legislation will struggle with assigning a price to these indirect costs for years, and honestly, neither side will ever be able to prove its case.

Biblical wisdom should be used, and we should err on the side of caution, which means taking the high road—the renewable route. We don't have to wait for politicians and government to mandate carbon legislation; we can choose our own path, to follow renewable alternative technologies or not.

The biggest challenge for many renewable technologies used in electrical power generation is the intermittency factor. The wind doesn't blow on demand whenever and wherever you want it, and the sun doesn't shine at night or on cloudy days. Tides go in and out only twice a day, and biomass is not available year-round in many locations. Hydropower can be affected by drought and timing of rainfall, as can geothermal. Compounding the difficulty, electricity does not store well in bulk. AC electricity doesn't store well at all, and DC rechargeable batteries cannot provide utility-scale megawatts of power long enough, without lengthy recharging.

But when coupled with energy storage technology, renewable energy makes sense. They must provide consistent stable output to garner a larger share of power generation. The problem again becomes the cost. However, as the linking of carbon emissions to environmental costs progresses, the cost gap between traditional energy technologies and alternative technologies will close.

The renewable energy route, although unrefined like a country road, can someday be paved with gold. Since renewable energy works well on a small scale, distributed generation can relieve some of the power concerns. Distributed generation is the opposite of the traditional centralized power generation model. Generating electrical power and heat at the point of use is the method behind distributed generation.

Traditional methods centralize power generation and fuel refining, and often transport it long distances to get to the end user. However, with drastically increasing transportation costs, especially for vehicle fuels, the distributed generation method becomes more economical and efficient. In addition, distributed generation gives individuals and small local communities energy independence and flexibility. Plus, distributed generation reduces traffic on the roads of our energy analogy by lightening the load on the centralized energy infrastructure. Laws enacted in many states now allow families to sell extra power back to the electric grid, providing extra revenue for the family. This benefit of distributed generation—*selling* power to the utility company instead of *buying* power—would really catch on with entrepreneurial Americans, given the right incentive pricing and simplified electrical interconnect regulations. Several large states, such as California and Texas, have implemented attractive net metering regulations that consumers are slowly embracing as they become more familiar with them.

In some ways, centralized power generation and distributed generation mirror our church and home relationship with God. We need to congregate with fellow believers and leaders in the *church*, which is like receiving power from a centralized power plant. Likewise, we need to develop a personal relationship with God in our *home*, which is like our distributed generation energy source. As we build our everyday relationship with God in our *home*, we learn to minister to others, to spread God's word, without relying on the *church* to transmit all the power of God's message. But just as the Bible teaches we need both the fellowship of believers and a personal relationship with God, we also need both central and distributed power generation for a healthy life.

RENEWABLE ROUTE POTHOLES

On the fringe, universities and corporations have been chasing the dream of renewable energy for years. However, they encountered many potholes along the way, including: the inertia of the massive fossil fuel industry, inconsistent or non-existent government support, an entrenched auto industry, and consumers who for the most part didn't care. Other potholes include: the amount of extra land that wind and solar power require compared to incumbent technologies of similar size; proper environmental site location, since renewable technologies have an impact, too; NIMBY (not in my back yard); and scalability of renewable energy to traditional megawatt-sized central generation infrastructure.

CRYING WOLF?

Perhaps the biggest pothole in accepting renewable energy is associated with crude oil. Pessimists of the energy crisis have heard the ill-fated "cry wolf" calls before, from the first big oil-related crunch brought about by the 1973 oil embargo, to recent crises, such as Hurricanes Katrina and Rita. Given time, supply has always stabilized and price settled to a normal level. The average American has trouble believing there are truly oil supply problems brewing, because we have been "wolfed" too many times before. Have we been lulled to a sense of complacency when addressing energy-related sustainability, especially with oil and gasoline? Cries to move away from unsustainable energy and environmental degradation often die as fast as the attention span of a two-year-old.

"But *this* time it will be different!" many exclaim in environmentalist circles. "We need sustainable, renewable energy now!"

I can hear the evening news reporter now. "Al Gore cries out, 'Environmental disaster!' exclusive at 10 P.M." Is this the time that the wolf actually comes? Is Al Gore bumbling a statement like his infamous "inventing the internet" misquote, or speaking "an inconvenient truth?" The point centers not on Al Gore's truthfulness as an environmental prophet, but on the truth that we should not rush to judgment on a subject, no matter what we think, favorably or unfavorably, about the messenger. We would be wise to inconvenience ourselves, at least long enough to see whether it's a wolf or a friendly dog coming down the road. It's important to remember how the fairy tale ends; the wolf devours the flock as the complacent townspeople ignore the boy's cries for help.

The wolf in our world manifests itself as unsustainable energy resources. We have been told before that the wolf was at the door, that we were at the tipping point of unsustainable energy production, only to find it wasn't true. The truth is our current energy infrastructure is not sustainable. We can deny it all we want, but it won't make fossil fuel energy supplies last one day longer. Although many have cried wolf before, the wise will heed the warning signs and switch to renewable energy, building an off-ramp from fossil fuels to connect to the renewable route. The road connecting fossil fuel and renewable energy might be a rocky one, but it's a critical off-ramp to build.

Renewable energy provides 2.5% of total U.S. electrical power generation (excluding 6% hydro). For the transportation sector of the economy, renewable energy provides 2% of the total fuel.[11] These small percentages testify to the depth of our complacency regarding energy. We are stuck on unsustainable energy and the call for renewable energy falls on deaf ears. If the wolf came today, we would be eaten alive.

The renewable route offers many environmental benefits, energy diversity benefits, and energy security benefits. But the deal-breaker so far has been a perceived lack of economic benefits. Unfortunately, it is wrongly tagged as commercially unviable, which may have been true thirty years ago, but today becomes a question of what we value. If we value environmental benefits, energy diversity benefits, and energy security benefits, we had better get past the traditional economic views of renewable and alternative energy. This remains especially true for motive vehicular power, which is 96% fossil fuel driven. Regardless of history, the difference this time is two-fold: one, the leveling or reduction of energy resources, combined with population growth and middle class growth in developing countries, is legitimately straining energy supplies. Two, as a result of reason one, the cost of traditional energy supply, whether it be oil or gas or other, is increasingly volatile, making the cost of "fuel-free" renewable energy competitive. After all, God doesn't charge us any more for the sun, the wind, the tides, or the waves than he did Adam and Eve—they're free!

Nuclear Street

On-ramps to Nuclear Street are being built in the U.S. after many years of self-imposed exile. Many see nuclear power as an avenue to cleaner air and diversified electrical generation; therefore, new nuclear plant construction in the United States seems likely. The huge capital cost and difficult financing make it challenging, but the operational costs are cheaper than almost every other source of power generation. Even though the operational costs are lower, the social, financial, and bureaucratic uncertainty keeps many new nuclear plant designs on the side of the road. New applications for nuclear power plants first look to build new reactors at existing sites in order to reduce red tape as much as possible.

Even though new nukes have been at a standstill, just over 100 nuclear reactors operate in the United States at 66 sites, producing an incredible 20% of the nation's electrical power. Part of the story has been the increase in power output capacity from the existing plants and the increase in plant availability, a measure of how reliable the plant is at producing electricity. The average capacity factor is approximately 90% compared to less than 60% for nuclear energy in the 1980s.[12]

With nuclear energy, the biggest pothole in the street is safety—or the perception that it is not safe. The United States market for new nuclear power plants has lagged behind the rest of the world since the mishap at the Three Mile Island site in 1979, and no new plants have been designed and built in the United States since, although several in progress were completed.

Although the safety systems worked at Three Mile Island and no release of radiation occurred to the public, the media outrage brought nuclear expansion in the United States to complete gridlock. No one died, but if you ask the average person on the street corner if people died in the Three Mile Island "disaster," I bet many would say human lives were lost.

Unfortunately, reactor designs elsewhere in the world sometimes lack the safety features and operational safeguards found in Western Europe and America. The 1986 Chernobyl nuclear disaster in Russia remains a haunting reminder of what can happen when nuclear power goes wrong. The estimates vary widely on the number of deaths related to the Chernobyl meltdown. The World Health Organization (WHO) cites only fifty direct deaths, while Greenpeace declares the number ultimately will be close to 100,000 deaths.

Obviously, someone will be wrong, but an undeniable truth rings loud and clear—that nuclear accidents can affect lives for generations. Do not take nuclear safety for granted; however, do not let it paralyze you in fear either. Compared to fossil fuel mining or drilling, and air- and water-pollution risks from burning coal and other fossil fuels, the safety record of nuclear is quite good. Moreover, the nuclear waste generated does not pose as much of a threat as many fear, since the quantity of nuclear waste is very low compared to the energy output. Much of the spent nuclear fuel, if reprocessed, can power reactors again, which is a procedure not presently used stateside very much.

The other potential pothole on Nuclear Street casts an ominous shadow on safety—the dangerous threat of terrorism. But *fear* of nuclear power plant terrorism, at least in the United States, is much greater than the likelihood of an actual attack. The greater terror concerns by far should be from nuclear weapons, nuclear reactors in lesser protected countries, and nuclear technology of any kind in countries such as Iran and North Korea.

However, the true danger looms because the growing public comfort and support of nuclear power gridlocks in a millisecond if there is just one accident. It doesn't matter if that is a fair assessment of the safety of the technology and the industry. The truth remains that some forms of radioactive waste take hundreds to thousands of years to decay to safe levels. Its fallout affects humans, the environment, and the economy for years to come. That scares many into fighting nuclear energy staunchly. A certain percentage of the population will never get over that hurdle. If a nuclear incident happens, by accident or by terrorism, the mere *perception* of a problem could be enough to drive a large percentage of the population to abandon support of nuclear energy.

Much like the Fossil Fuel Freeway, Nuclear Street continues to provide power, at least in the near future, since it is unwise to abandon existing energy infrastructure too quickly. However, by the turn of the next century, if not sooner, careful consideration should be given to the continued use of nuclear fission technology. The world would be a safer place if Nuclear Street closed to all traffic forever. However, the energy solutions are too far down the road to decommission nuclear fission reactors anytime soon.

Until we solve the mysteries of cold fusion and fly around in converted DeLorean time machines (like in the movie *Back to the Future*, with our flux capacitor and a "Mr. Fusion" unit attached), we are stuck with "Mr. Fission" when it comes to nuclear power. Nuclear fusion and nuclear fission are kind of like heaven and earth. God demonstrates tremendous power by putting atoms together (fusion), while humans tear atoms apart to demonstrate our power (fission). God's ways are always better. Nevertheless, on Nuclear Street, drivers emphasize safety with nuclear fission, and usually deliver. It's the threat of a monster wreck that worries drivers.

Nuclear streets are like main avenues—big, wide, well-traveled, but with lots of stoplights and traffic snarls that could pop up quickly, bringing traffic to a standstill. Nuclear power is desirable because of the amount of energy it produces per pound of fuel spent and because it provides megawatts of electrical power within an extremely small land area. But miles of red tape, inaction due to political and legalistic gridlock, and the threat of big disaster could quickly bring electron traffic to a standstill if a nuclear incident happened. And that's a big "if," because the facts surrounding nuclear power generation in the

United States support its remarkable safety. But like a herd of elephants on 5th Avenue in New York City, nuclear energy can disrupt the electrical power system like no other technology.

The "fission" lanes of Nuclear Street someday will close permanently, if we master nuclear fusion. In the meantime, the planning of new on-ramps to Nuclear Street has begun. The promising avenues of nuclear energy are Generation IV technologies, which link higher temperature processes beneficial not just for electricity generation, but also hydrogen generation. High-temperature electrolysis thermal cracking of water generates hydrogen scale much more efficiently than traditional electrolysis. Another Generation IV regenerative process that holds much promise is thermal-chemical cracking. The sulfur-iodine process is one of the front-runners for this mode of hydrogen generation.

The nuclear industry has generated electricity and, surprisingly, supporters who consider it a new green solution that will rescue us from fossil fuels. No less than the founder of Greenpeace, Patrick Moore, is a member of the green nuke camp. Considering also that large-scale hydrogen generation without the use of fossil fuel is possible with nuclear power, it warrants consideration as a green solution and a key to diversify our energy infrastructure. And God willing, nuclear fusion may someday eliminate the radioactive nuclear waste potholes of nuclear fission.

Bio-fuel Boulevard

I heard about a car that ran on beer from my friend, Dave. I immediately thought, *What an interesting bio-fuel (alcohol) usage, but a waste of good beer.* Then Dave told me the catch: you promise the driver beer, and he'll drive you in his car wherever you want to go.

Kidding aside, bio-fuels such as ethanol boomed in popularity recently, although a backlash has occurred, due in part to the phenomenal expansion that grew too quickly to sustain a healthy economic foothold. For a while, ethanol projects were springing up like Starbucks locations. This has snapped back to a reasonable level. Given the ethanol gains from increasing growth of government incentive programs and mandates, and considering the direction of oil markets, the ethanol trend will probably continue.

However, detractors of corn-based ethanol like to point out the potholes in Bio-fuel Boulevard. The rapid growth of ethanol has created a battle of land usage: food crops versus energy crops. Either way, farmers enjoy the increased prices for their products, although fuel and fertilizer cost increases have eaten away most of those profits. Another pothole in Bio-fuel Boulevard is that the energy content per gallon is less than the equivalent amount of gasoline. Experts argue over the size of this energy penalty. Is it about 3%, or 15% for the standard 10 % ethanol (E10) blend? Or some other percent? The 85% ethanol blend (E85) have even greater energy content losses. Either way, many consumers hate the perception of paying for something less than what they expect. Detractors like to harp on the fact that corn-based ethanol is subsidized at 51 cents per gallon, while high protective tariffs keep cheaper ethanol from Brazil and other places from reaching consumers in the United States and reducing our fuel costs.

Ethanol has some driving factors creating on-ramps to the Bio-fuel Boulevard. For one, it's the major oxygenate used in reformulated gasoline since the environmental and legal downfall of Methyl Tertiary Butyl Ether (MTBE). Secondly, the most important benefit by far is the fact that it displaces demand for oil, especially foreign oil. I would much rather pay a farmer in Iowa for fuel than a sheik in the Middle East or Chavez down in Venezuela.

Many complain that corn-based ethanol is carbon neutral at best. The carbon absorbed in the growing process is largely offset by the fossil fuels burned to plant and harvest the corn, and the large amounts of fossil fuels burned in power plants to power the distilleries. The process uses large amounts of water in areas already facing low aquifer levels and water restriction (the Midwest), where most of the bio-fuel crops are grown. Fuel crops displace land that could be used to plant food crops, preparing what could be the groundwork for a future dilemma of food versus fuel. Ranchers and dairy farmers don't particularly care for corn ethanol because it raises cattle feed prices.

Despite corn ethanol's challenges, it does displace foreign oil imports and puts more money in the hands of farmers and American businesses.

It also paves the way for new alternative fuels, getting the public accustomed to fuels other than gasoline. It is important to prove to the public that other fuel choices work, if the addictive cycle of oil is to be broken.

Part of the next wave of ethanol technology, cellulosic ethanol, will relieve some of the corn-based ethanol troubles. Cellulosic ethanol encompasses anything grown: from wood waste, to fast growing grasses that require little water, such as switchgrass. The challenge surrounding these materials entails the difficult and costly conversion to ethanol.

I've heard it asked on more than one occasion, "If Brazil can switch to ethanol, why can't we?" Brazil does run the majority of its flexible fuel autos on ethanol made from sugarcane. And sugarcane is much easier to convert to ethanol, since it's much higher in sugar than corn, therefore much of the enzymatic process in producing ethanol is bypassed. However, sugar cane doesn't grow well in large areas of the United States, and Brazil clears valuable rainforest to plant the energy crop, wiping out part of the environmental benefit. Brazil also has only 19.4 million vehicles, compared to over 230 million for the U.S.,[13] making it much easier to convert quickly to a new source. The Brazilian government also established a framework to support the growth of sugarcane alcohol. The origins of their alternative fuel transfer stem back to the 1970 oil shortages. Today, most of the newer vehicles are flex fuel, capable of burning either gasoline or sugarcane blend.

Bio-diesel is another option that has gained strength, with a fast growing replacement of petroleum-based diesel. Diesel fuel in general has cleaned up its act in recent years, with new low sulfur fuel requirements, and the fact that diesel engines are the most efficient of the combustion engines. Bio-diesel takes an additional cleanliness advantage over the low sulfur diesels in the way it is created: taking waste cooking oil and other fuel feed stocks, such as canola (called rapeseed in Europe), and making carbon neutral fuel.

Unfortunately though, it shares several of the same potholes as ethanol, such as limited feedstock. Plant-based bio-diesels have a simpler and more homogenous product, but like ethanol, the feedstock has to be close to the processing plant to be economically feasible. Animal-based oils used to create bio-diesel are more complex and harder to process, plus they tend to coagulate easier in cold weather, making vehicles harder to start. Used cooking oil is a cheap (and sometimes free) waste product that many restaurants are happy to supply. There are only so many fast food burger joints though, so this "free" supply could dry up quickly.

To wrap it up like a Big Mac to go, bio-fuels are so enticing because they are liquids, which we are accustomed to putting in our gas tanks. Other fuels that could be used for transportation, like natural gas and hydrogen gas, are not nearly as familiar. Even plug-in battery-powered hybrids would require an adjustment. However, many believe that bio-fuels are not the final answer to our petroleum problem, since they take too much land and energy to process the fuel. But the Bio-fuel Boulevard will be part of the solution, because it reduces foreign oil imports, is almost carbon neutral, and its liquid form is easier to transition to than other fuels. So what if it doesn't answer all the questions; at least it provides an off-ramp from Fossil Fuel Interstate to other roads.

Electric Service Road

Few roads parallel the Hydrogen Highway closer than electricity, but somehow it seems to come up short of the final destination. Electricity functions more as a service road than a main highway. The electric service road goes the right general direction (paralleling the Hydrogen Highway), but it doesn't have the capacity, continuity, and high speed limits that the highway possesses.

Early in the development of the automobile, it wasn't certain which would be the dominant technology to drive the vehicle: the internal combustion engine (ICE), the electric motor, or the diesel engine. The internal combustion engine won the battle for market share, and perhaps succeeded too well, since it helped drive us to this dead-end alley of fossil fuel dependence. But the other technologies are still alive. In California, a law was passed requiring Zero Emission Vehicles (ZEVs) that led directly to the EV1 program, the first modern commercial test of all-electric plug-in vehicles. The program succeeded in many regards, including the fact that many of the drivers loved the vehicles. Nevertheless, the project died and, as detailed in the movie, *Who Killed the Electric Car*, California back-pedaled to the ICE.

One of the main reasons for canceling the EV1 program was that battery capacity could not match the equivalent "tank of gas" mileage of approximately 300 miles. The recharge times were also claimed by some to be too long. But many saw these as weak excuses to cancel the program, since the hurdles could have been cleared. For one, about 78% of personal automobile mileage is within forty miles of home, according to GM information.[14] Vehicles recharged at home could provide most of the driving needs.

The main potholes on the Electric Service Road lie with lithium-ion battery recharge times and longevity. These problems may fall to technology advances, but it is difficult to imagine them being resolved in a way better than hydrogen fuel cells. Batteries store energy internally. The quantity of energy and recharge time to restore that energy has physical limits that can only be minimized, not removed. Batteries will provide an on-ramp to merge on to the Hydrogen Highway.

Hydrogen Highway

Right now the highway is under construction, and like most big construction jobs, it's taking longer and costing more than people want. However, when it's done, it will be awesome and almost everyone will be touched by it.

Some say hydrogen is too dangerous a road to follow, too expensive, too uncertain. Don't we have other roads, like Route 66, to get us across the country? Like building the Interstate-70 Eisenhower Tunnel through 1.7 miles of mountain in Colorado, building the hydrogen highway won't be easy. But it's worth it.

The virtues of hydrogen discussed in detail in Chapter 6 provide much of the background for this section. But to recap the benefits, hydrogen is plentiful and available from many feedstocks, including water and fossil fuels. If renewable or nuclear sources of energy are used to generate the hydrogen, no carbon emissions or air pollution occur. Hydrogen burns clean without carbon dioxide emissions, and fuel cells utilize hydrogen that's even cleaner, with pure water vapor as the primary emission. Unlike all other fuels, hydrogen is not a pollutant if leaked. It is not a greenhouse gas, and dissipates quickly if leaked.

No hazmat cleanup is required, except for ventilating the escaped gas. In industry, hydrogen has demonstrated an exceptional safety record, a fact most people don't know. Hydrogen stores energy more effectively in bulk than batteries, plus the hydrogen refuels much faster than batteries can recharge. Overall, no other energy source or carrier can provide the volume, cleanliness, storability, and sustainability that hydrogen can provide.

If hydrogen holds all the answers, what keeps the Hydrogen Highway from progressing? The potholes of the Hydrogen Highway are primarily a few engineering breakthroughs and cost. Both are conquerable.

The engineering breakthroughs center around fuel cell durability, hydrogen storage challenges, and overcoming hydrogen-generating inefficiencies, which are not as significant as they appear on the surface. The next section tackles the efficiency dilemma.

Cost is the second concern of the Hydrogen Highway. Right now, the cost is prohibitive, but that comparison hinges on the fossil fuel prices that dominate the market today. As the sustainability and environmental issues surrounding fossil fuels become more prevalent, the costs of hydrogen will quickly become more competitive—faster than many predict. The costs of hydrogen infrastructure are manageable if we use distributed generation of hydrogen to jumpstart the hydrogen economy. From there, the larger central distribution infrastructure can develop as needed to support mass marketing of hydrogen. The high cost of fuel cells plummets as economies of scale and mass production take hold.

Some say fuel cell durability is in question and unproven on a commercial scale. Scientific improvements in durability and power output are advancing rapidly. Many companies are racing to solve these issues, and I'm confident they will succeed. In the meantime, the hydrogen economy can move forward by combusting hydrogen in modified internal combustion engines, or even getting consumers accustomed to gaseous fuels by using natural gas as an intermediate step. Using natural gas for transportation fuel is one of the measures T. Boone Pickens sees as a stopgap measure also, until sustainable alternatives become established.

The only other pothole on the Hydrogen Highway is the perceived safety issue. Yes, hydrogen is explosive and flammable, but so are many other fuels, such gasoline, propane, and natural gas. The public deals with these all the time, and familiarity breeds acceptance. Hydrogen is safe if handled with the right precautions and safeguards. In many respects, hydrogen is better: it dissipates quicker, and does not pollute. The unresolved issue that requires caution is the fact that hydrogen is odorless and burns with an almost invisible flame. That presents a more difficult technical challenge, since many fuel cell types require very pure hydrogen. Because of the hydrogen purity necessities, odorants and colorants are difficult to add. This challenge is not insurmountable either, and I'm sure this hurdle will be cleared.

Efficiency Overrating

Energy efficiency is often called "the fifth fuel," although it really isn't a fuel at all. It's a play on words since increasing efficiency "saves" fuel, no matter the particular energy source. Efficiency is good to have, but in many ways it is overrated. There is no practical way to conserve our way out of our energy dilemma, but we can lessen the impact in the short term, and build a foundation for supply side changes.

Do not get me wrong, energy efficiency improvements are the low hanging fruit, the most expedient way for us to reduce our present energy usage, and we should pursue that fully. However, as shown on page 163, a strong argument exists for focusing on the renewable generation of hydrogen for fuel cell vehicles (FCV) and not worrying about total energy usage. The present combustion vehicle (CV) uses more total energy, is less efficient, not to mention it consumes the most oil, and emits the most greenhouse gas. Reductions of oil usage and greenhouse gas emissions are ultimately more important.

People talk of the inefficiency in creating hydrogen, and all the energy that must be expended to generate hydrogen molecules. However, people don't think twice about the inefficiency of creating electricity from start to finish. The electricity arriving at the normal house outlet often contains only a third of the energy content that originated in the coal from which it started. Factor in the incandescent bulb inefficiencies of a typical house lamp plugged into that outlet, and you're at about 5% of the energy content of that coal. Factor in the energy expended to mine the coal and the efficiency drops to near

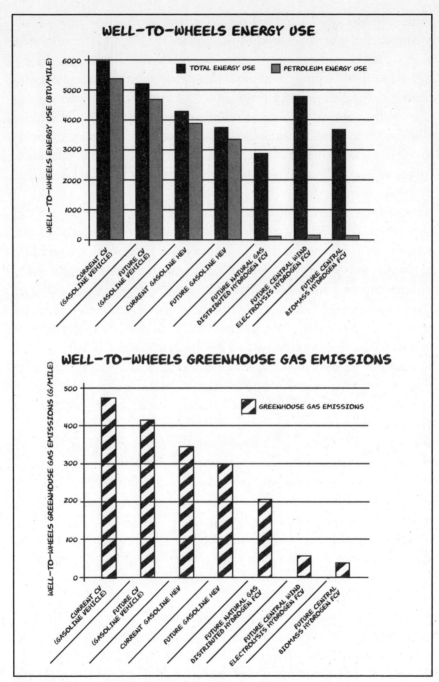

Figure 8.2—Well-to-Wheels Energy Use & Greenhouse Gas Emissions. (Courtesy of U.S. DOE, "Hydrogen & Our Energy Future")

zero. Tally the environmental pollution and greenhouse gas cleanup inefficiencies, which are hard to put a price tag on, and the story gets even worse.

It is also misleading sometimes that energy efficiencies are given from a start point midway through the process. For example, Energy Star-rated compact fluorescent bulbs are typically up to 75% more efficient than incandescent bulbs,[15] but that is from a start point of the light socket. This gives a deceptive picture of the overall efficiency. All the efficiency losses in getting the energy to the light socket are not counted.

Put into perspective, the efficiency of hydrogen generation and use in fuel cells is not so bad after all. The important consideration is not net efficiency, but net benefit. By net benefit, I propose we should be looking at many angles: economics, efficiency, environmental impact, societal impact, national security, and most important, God's will.

Likewise, the "well-to-wheels" efficiency of the gasoline cycle is just as bad. If you consider the efficiency of the oil extraction, transportation, refining, and the efficiency of gasoline engine, you have an overall efficiency of approximately 14%.[16] If you add in intangible factors, such as the costs of environment cleanup and a portion of the cost to national security, the efficiency is even worse. Here's an easy-to-remember, but sad rule of thumb: if the vehicle is 10% efficient, and at most 10% of the vehicle's total weight is passengers, then only 1% of the fuel consumed moves the passenger payload. In other words, 1% of the energy in your gas tank moves your buttock from point A to point B, while 99% is wasted. We won't get into how the size of your derriere affects fuel efficiency. Amory Lovins, leader of the Rocky Mountain Institute, presents very detailed information in his book, *Winning the Oil Endgame,* but confirms the disturbing 1% number. This means a whopping 99% of gasoline in the tank is wasted to heat and inefficiency.[17]

Let's compare the internal combustion engine to fuel cells. The fuel cell itself is much more efficient. Depending on whom you talk to, the average gasoline engine is about 10% efficient on the low end and 18% efficient on the high end. The average fuel cell for automotive purposes is anticipated to be at least 30%. The internal combustion engine has been developed over the last 100 years, leaving less opportunity for refinements in technology. The fuel cell, on the other hand, is new to commercial development and more likely to have efficiency improvements as its technology matures.

We know that it is important to efficiently use our fossil fuel resources, since it is something we can control until cleaner energy sources are dominant. The big picture with efficiency is often missed. The cleanliness of the source, how we develop it, and how we use it is ultimately more important than efficiency itself. As a case in point, notice that God created the earth with great inefficiency also. If you look at the efficiency cycle from sunlight to photosynthesis, there is a lot of wasted energy. If it is good enough for God to design with such inefficiency, maybe some of our preoccupation with it is misplaced. For example, the earth receives only a tiny fraction of the energy the sun gives off. Did God design the sun and our solar system to dissipate excessive power for a reason? We don't know why, but we do see the results and can measure the average amount of energy that reaches the earth's surface. It is about 2 billionths of the total energy of the sun.[18] Not very efficient, but if we received even 1% more of the sun's total energy, the earth would be a molten ball of rock.

The efficiency with which plants convert sunlight to sugar in the photosynthesis process is also only 0.023%.[19] I propose that if efficiency was that important, God would have made it more efficient. Maybe the lesson lies not with overall efficiency, but with how the energy is used.

There is also a catch-22 with efficiency. With higher efficiency, operation costs drop. As operating costs remain low, the easier it is to become complacent, using more fuel since the fuel costs are lower. Economists call this a backlash effect, increasing the amount of energy usage. Efficiency can never do more than widen the road. Taking a road and widening it makes it more efficient. More cars get through, quicker, with less traffic jams, but it does nothing to change the direction of the road or the destination. What it gives is time to build an off-ramp.

Energy efficiency may help us make better use of our natural resources, extending the amount of time they're available and temporarily reducing our dependence on imported oil. Improving efficiency does nothing to change the course of fuel choice. In fact, it can extend the duration of that fuel choice. If unaccompanied by an attempt to transition the massive energy infrastructure to better alternatives, improvement in efficiency is time wasted. Efficiency cannot do anything more than delay the need for imports and the inevitable end of fossil fuels. World population is too big for us to "conserve and efficiency improve" ourselves out of fossil fuel dependence, and going back to horse and buggy days is not an option. Energy efficiency buys time. Using that time wisely to build roads to a sustainable and alternative energy future remains the best use of efficiency and conservation measures.

It's important to consider efficiency when enacting lifestyles to follow God's will, but it's not the road to energy freedom. If a self-sustaining, fully renewable hydrogen economy develops, then efficiency won't matter much at all—you'll just pay for the energy you use (or waste), but the environment, and energy security won't be negatively affected. It becomes a matter of personal economics; wasting energy harms your own pocketbook, not others, or God's creation. The days when we don't need to worry about the amount we waste are nowhere near, but conscientious consumers can dream of that freedom from harmful waste. Until then, we should pursue efficiency earnestly in conjunction with our present day energy infrastructure, but fully realize the real answers reside on the supply side of the energy equation.

The Jet Stream

Some really smart people don't want to follow any of the roads we mentioned. They plan to fly, instead. They believe that the end of times is inevitable and God will restore everything, including the planet, when Christ returns. So why build new roads? Although God surely will restore the earth, and do so beyond anything we could possibly image, we're still required in Scripture to care for the earth until Christ does return. *Christianity Today* explains it this way, for those who have an eschatological (end of times) view:

> . . . An eschatological perspective helps us save nature for God's sake, not just for our own benefit. Care for the natural world is not just about cost-benefit analysis for human

welfare, though that must always be done. But if God has a plan for this natural world, has a bright future for it, we do not always need to see the benefit for ourselves before acting to preserve the natural order. It should be enough for us that this is part of God's vision for the future and a carrier of his promises.[20]

As Christians, we know heaven is our eternal home. Nevertheless, we still must live our lives here. And since we have been saved, we are called to do good works, which includes caring for God's creation. Jesus tells us about the end of times in the Bible and teaches us to be prepared. Many books have been written about eschatology, but an important thing to remember is taught by Jesus in the parable of the ten virgins (Matt. 25:1–13). Half the wedding party runs out of oil (deja vu all over again) for their lamps and tries to purchase some too late. The guests who did not prepare well were locked out of the wedding when the groom arrived. Jesus taught that we don't know when the end of times will be, but we need to be prepared. Caring for the earth is part of that preparation. The Bridegroom comes at midnight, during the darkest time. We likewise should still be prepared during these dark times of economic and energy crisis.

Here's a modern day parable: A father tells his two children to clean their rooms. The first one procrastinates and doesn't clean the room, and even trashes it more. The second child keeps his room tidy all along. When the father returns unexpectedly, with whom do you think the father is pleased? The father still loves both children, but demands that his commands be obeyed. You can bet there will be consequences for that first child. I would much prefer to be the child who actively followed the father's command.

Seeking clean energy helps us to be prepared because it leads to green stewardship. In other words, we have reason to continue to build roads, even if we plan to fly. After all, one needs roads to get to the airport.

CHAPTER 9

DRIVERS ON THE ROAD TO FREEDOM

From everyone who has been given much, much will be demanded; and from the one who has been entrusted with much, much more will be asked.

—Luke 12:48

Have you ever read a personality profile? They provide fascinating insights into personal tendencies. Simply put, there are four main personality categories: *analytical, driver, amiable, and expressive.* It's important for people to know and understand different traits in other people to learn how to more effectively interact with each other. For our purposes, we're focusing on *drivers.* The personality traits of a driver reveal a person who leads or tells others what to do, makes decisions, and focuses on the task. One can see why this personality profile description originated from drivers of automobiles. A driver of a vehicle steers the car, controls the speed, makes decisions based on the traffic laws, and chooses the roads required to get to the desired destination. It's no wonder Americans love to drive. The freedom and control are exhilarating.

The rest of the world, especially China and India, continues to expand the numbers of elated drivers every year. China and India added millions of private car motorists to the road in 2007, with the trend expected to grow 10% to 15% more per year.[1] By comparison, the United States market is saturated; personal auto sales shrank slightly in 2006. The size of a country's total population is the key to the growth. For total population, the United States averages 0.75 private automobiles per person,[2] while China averages 0.09.[3] India's percentages are similar to China's. The potential growth in the number of drivers results from the strong growth of the middle class in China and India.[4] This conveys exciting news, but also presents danger to the ecological and economic balance of the planet if the status quo of oil-powered transportation continues. We need new, non-petroleum cars for drivers in the United States and across the globe.

However, not everyone can or should be a driver: some are too young or too old, and some are physically or mentally impaired. Some abuse alcohol and drugs and lose the privilege to drive. Sometimes able-bodied drivers are restricted to the role of passenger since only one person can drive at a time. And the role of passenger doesn't sit well with everyone. Those passengers who think they should still have the wheel often verbalize their opinions and become backseat drivers.

The bottom line is that driving is a privilege, not a right. It is one privilege Americans take for granted, as if driving is an entitlement. For example, some rude drivers assume they own the road, taking their need for control too far. The least courteous drivers fly into fits of road rage, too often resulting in accidents, physical altercations, and legal trouble. That entitlement attitude remains deeply imbedded in American culture, business, government, and even sometimes in our religion. This perceived American birthright—to do whatever we please to support our oil habit—causes the country much ill-will internationally. The United States also gives aid and support more than any other nation, but we erode that international goodwill when we expect on-demand crude oil and resources whenever and from wherever we want. We need to find another way to support our driving habits that doesn't require petroleum.

Consider for a moment if the roles of Iraq and the United States were reversed. Pretend a rich and powerful Iraq had used their domestic supplies of petroleum and poor, undeveloped tribal Texas oil was in abundance. Also pretend Sunni and Shia Muslims were living peacefully together and with the world, while Catholic and Protestant

Christians conducted a civil war in America. What if Iraq needed a geo-politically stable United States region (call it the Middle West) to ensure oil supply to the Iraqi superpower economy? How would the United States react if the Iraqi military set up shop in Texas to protect us from Christian religious wars and liberate us? It's hard to imagine a peaceful and gracious response from Texans for the Iraqi liberators. The last time a supposed liberating force came to Texas, they ended up remembering a little place called the Alamo.

This far-fetched scenario sheds light on something that Americans often assume—that other nations want our involvement in their affairs. Sometimes they do, sometimes they don't. In many ways, we're acting like England did with the American colonies just before the birth of our nation. We somehow have adopted the empire mentality that bringing Western-style democracy to the Middle East will lead to freedom. More likely the opposite will happen. Democracy will lead to Islamic theocracies and dictatorships, at least as long as the oil holds out.

In the meantime, many Americans, at least subconsciously, act as if we are entitled to free market access to oil and material resources, wherever they reside on the globe. It's the inevitable outcome of being a world superpower for two generations, having fought all our wars on foreign soil since the American Civil War seven score and four years ago. Even if we pay a fair market price for the black gold, we know that is not enough to sustain our lifestyle. We also use questionable diplomacy with nations like Saudi Arabia and our military might to guarantee that access for our national security.

On the other hand, as the world's leading superpower, we do have a responsibility to protect and stabilize regions around the globe. Without cheap access to oil, in the present world energy configuration, it's not possible to fulfill our perceived role of "world protector." I am convinced the world is a better place with America in that role than any other country on the planet, but it also takes its toll on us.

Regardless of justification arguments, the fact is, as world oil supplies tighten, it would be unwise to expect foreign countries and companies to supply oil as usual. At a minimum, the oil-rich nations will expect a severely high premium. People and nations will protect their own first, likely leading to resource skirmishes and eventually resource wars. That's why developing alternative energy should demand such attention.

The United States commands the driver's seat now, but the backseat drivers on the planet look to take the wheel the first chance they get. Right now, the United States uses oil resources faster than the developing countries can expand their economies. Is the goal for the Western world to use up all the oil resources before developing economies grow their middle class to challenge our uber-consumer lifestyle? It's like speeding to the gas station because you're running low on gas. It exacerbates the fuel shortage by decreasing the fuel efficiency by speeding—not to mention the added risk of a speeding ticket or accident. Maybe this shortsighted view is marginally acceptable to the present and next generation, but where does it leave our grandchildren and the poor of the world?

When you look at it, America's expectation to always command the wheel mirrors secular disregard for God's ownership of all creation. We're merely stewards of God's creation, yet most of the time we act like the rude driver who thinks he owns the road. We need to drive the use of God-given resources safely and carefully, following the rules of the road (God's commands) and develop the right roads (God's will) to lead the rest of humanity to God's love and salvation. It's not unreasonable to expect energy and environmental paths to play a part in God's mission here on earth. The question remains, will the various drivers on the energy roads act as good drivers, or will they spinout and wreck in fits of road rage in a fifty-car pileup?

In the last chapter, we explored the roads to energy freedom. In this chapter, we will look at various drivers of our energy choices. For example, we'll examine how big government acts as a construction crane, powerful but only as effective as the party behind the controls, and how regional government acts as a city bus, responding to the varied needs of its passengers. Other drivers include: businesses, who serve their customers, much like the driver of an ice cream truck; consumers, who sadly have taken a role as truck drivers of cheap trinkets; Christians, who like a school-bus driver, should seek to care for the safety of others; and last but certainly not least, women drivers, who control considerably more than the minivan on the way to soccer games. The drivers are ready; time to hit the road.

The utilitarian purpose of the road is to serve the driver. The driver makes the decisions: where to go, what road to take. If the roads don't connect smoothly to where the driver wants to go, new roads are designed and built. If roads contain potholes the size of Volkswagen Beetles, drivers pick a safer alternative route instead. Our energy problems cannot be fixed by constantly patching the potholes of failing road systems.

The same is true for our energy infrastructure. Leaders need to use wisdom and avoid the potholes of unsustainable energy sources. Wise leadership ensures smooth construction of the roadbed, which allows the building of new alternative energy infrastructure. Talk of building a new road is not enough. Will Rodgers said you can be on the right track but still get run over if you are not moving forward. It's time for our leaders to take the wheel and drive us where we need to go. We've been stagnant too long; it's time for us to demand that leadership. The potholes of the fossil fuel road have been neglected too long. Those roads are no longer effective and should only be patched. It's time for drivers to demand a new road—and see it through to completion.

Of all the drivers discussed in this chapter, Christians may not come to mind as the frontrunner. However, if the world's 2 billion Christians collectively wake up to the importance of energy in God's call to stewardship, God only knows what we could do. Christianity leads us to fulfill our inconvenient purpose: to honor God in everything that we do, including how we use energy. We should strive to uplift God's heavenly and earthly kingdoms in our daily faith-walk. No matter our station in life, profession, or gender, we should use those talents and time to honor God, serve the needs of the poor, and care for His planet. After all, it's all connected.

In fact, one should chuckle whenever hearing talk about "string theory," the scientific search for a unifying string that connects all energy and matter. Obviously, the common thread or "string" of our Creator's hand already exists in everything, whether the world recognizes it or not. Maybe science will recognize its folly someday and rename it "silly-string theory." Nevertheless, our Lord provides to all humanity common threads that bind us together, written on our hearts (Rom. 2:14–15). One example is the green stewardship that weaves its way into our personal behavior as politicians, citizens, producers, consumers, and even our gender as we respond to God's spiritual call. Look at the common thread of Christ's invisible hand in some of the usual, and unusual, places leadership may develop, and the importance of recognizing our Christian leadership is evident.

Political Drivers

It's been said before that science tells you what you *can* do, economics tells you what you *should* do, and politics tells you what you *will* do. My convictions tell me this adage is wrong, but it is still enlightening, showing the mixed-up order by which the world lives. If politics is the ultimate word on what we will do, then we're in a heap of trouble! Thankfully, hope resides in loftier places: "Find rest O my soul, in God alone, my hope comes from him" (Ps. 62:5).

Mountains of books, magazines, and talk shows scrutinize the political world and the economic world. Nobel prize winners, political activists, economic gurus, and other experts are all over the airwaves and the internet. The conclusions in this book may seem exceptionally simplistic when it comes to political-economical explanation: that's intentional. We complicate matters more than we should; a side effect of too many decisions based on what *we* think is best and not enough consideration of what *God* requests from us.

Big Government

Big government can serve as a catalyst, or a construction crane if you will, to effect great change. However, when it comes to driving the construction crane, we've put a lawyer (which most politicians are) in a construction worker's job to operate it. Finding lawyers who really understand the present-day energy system is difficult, and finding ones capable of doing the heavy lifting job to build a new energy system is even harder.

Voters decided to change course in the 2008 election and historically elected Barack Obama the 44th President of the United States. The new administration and congress must attempt to construct a new sustainable energy infrastructure amidst the most difficult times we've had since the Great Depression. Like the 23rd Psalm, we'll have to go "*through* the valley of the shadow of death"; we don't get to go over it, under it, or around it. One thing is for certain—this valley is deep, and the Obama team is going to have a hard time driving the big crane through. However, make no mistake, big government is on the move—like it or not.

Sadly, as we have discussed in previous chapters, we cannot rely solely on government to fix our energy problems. For example, one needs only to look at the tremendous growth in national debt, especially foreign-held debt. That is truly disturbing. The national debt (according to the U.S. Department of the Treasury, Bureau of the Public Debt), has grown at an incredible rate: at least $500 billion every year since 2003.[5] That equates to about $1.37 billion a day, or $15,800 a second. The stimulus packages and bailouts beginning in late 2008 are changing the national debt so quickly that the outrage of hitting 10 trillion dollars is a distant memory. In the media, the crux of the country's problem remains ignored. It all ties back to how we use our energy resources, the backbone of our economy. To many, it appears we're spinning out of control and our politicians are asleep at the wheel.

For example, pay-as-you-go budgetary rules elapsed in 2002, allowing government fiscal responsibility to fly out the window for both parties. Government deregulation of energy and fiscal markets resulted in an irresponsible romper-room atmosphere in too many business board-rooms. However, swinging the pendulum back to the day excessive regulation and bigger government could be just as damaging, pumping up our national debt even higher, at the expense of our future and our children's future. Whichever party looks beyond the short-term politics, economics and power struggles, and ahead to their God-given responsibilities, deserves to lead the country.

$700 BILLION AND $700 BILLION (REPEATEDLY)

In September 2008, the financial markets melted down. Deregulation on Wall Street in 1999 resulted in an atmosphere of greed that brought large investment banks, mortgage banks, and insurance companies crashing down. The president and congress went into reactionary mode in a struggle to put together a $700 billion dollar bailout package to save the financial markets and attempt to avert a full-fledged economic meltdown. There was plenty of blame to go around, including big government, for letting our ice cream truck drivers (Wall Street) drive the economy while drunk on greed. Wall Street, like any addict, refused to admit their problems until they had wrapped the ice cream truck around the flagpole on Capitol Hill.

Does that $700 billion number seem familiar? Lost in all the commotion of the $700 billion *one time only* bailout package for Wall Street, is the T. Boone Pickens Plan,[6] which claims that about $700 billion dollars *each year* will escape our economy if gasoline prices settle near $4.00 per gallon. Money to pay for energy imports flows into the hands of foreigners. In return for our $700 billion, we receive fuel that we literally burn to create power to drive our economy. Even then, there is no guarantee the power generated will produce anything productive; much of the power we generate is wasted.

The $700 billion burn of petroleum produces side effects: pollution, greenhouse gases, and continued addiction fed for another year. Then we do it all over again the next year, and the next, giving billions of dollars, filling the coffers of OPEC and others. Oil-exporting countries love to see us come with trucks of cash in tow, just like drug dealers love to see their favorite crack addict drive up with cash in hand. And their favorite addict is only determined by who has the most money; it's not true friendship. Simply put, it's not a caring and loving relationship, like God intends for our friendships. However, as noted in the last chapter, increasing domestic supply of oil (offshore and ANWR) only keeps our addicted drivers on the streets in better neighborhoods. It does nothing to get the oil-drunk driver off the street. The addicts should seek rehab, not suppliers closer to home. If we commit to a fuel cell economy soon enough, the addicts can sober up on good clean hydrogen and renew their driver's license in good standing.

The lack of a solid national energy policy has abetted our economic problems and national security problems. The $700 billion dollars for the financial market bailout will seem like the price tag for a bicycle compared to the potential dangers of ignoring energy and environmental problems. Politicians could support legislation to diversify our energy mix, promote alterative energy, and provide more distributed generation. However, they rarely get energy and environmental legislation passed until after the crisis is on top of us. As a case in point, the thirty-plus-year gap between energy crises when no national energy policy passed until the Energy Policy Act of 2005, comes to mind.

Exceptionally good long-term energy and environmental policy enactment cannot be rushed in time of crisis. The Energy Policy Act of 2005 is just a start; much more should be done, and done better. That appears to be the fire-drill mode we're stuck in. Politicians and voters alike need to prepare to combat long-term problems with long-term policy.

As an example of compromise, go ahead and lift federal bans on oil drilling in some offshore areas, but give zero tax breaks or government incentives to drill. If the economic reward is worth the risk, the oil majors will still take it—eventually. But more importantly, realize once and for all that answers do not lie with more oil. God's love is infinite; the supply of oil is not. The push for energy diversity and alternative energy deserves much greater attention.

POLITICAL OSTRICHES

I hate to see people shy away from the issue of green stewardship in life, in faith, and in politics because they perceive themselves as pragmatic, conservative Christian, or pro-business. For example, stereotypical republicans perceive all environmentalism and alternative energy legislation to be a liberal, secular, tree-hugger, anti-business, even anti-American. Any hint of democrat tied to an issue raises their defenses faster than Usain Bolt runs the 100-meter dash.

The same type of overreaction holds true for stereotypical democrats. The liberal, progressive ideology often paints all pro-business, fundamentalist Christians as willful, evil destroyers of the environment. Some see republicans as people who expressly seek to exploit the environment and fossil fuels for every ounce of personal gain. The hint of republican tied to an issue conjures visions of a Dick Cheney/Dubya Bush two-headed reptile like Godzilla burning everything in sight in Washington D.C. with its fiery breath. The democrats cringe at the thought of all the carbon dioxide produced from the reptile's rampage. Of course, this is an exaggeration. However, too many progressive liberals claim self-righteous justification from caricatures such as these. For both sides, refusing to work with the other does more harm than good.

The very goals of progress which both sides seek stagnate, due to their refusal to be inclusive and patient. Partisanship entrenches itself, refusing to work with the other group for fear of bending to their ideology. It sounds too much like a husband and wife arguing through a backseat driving episode. In both cases, you can decide who's behind the wheel.

No one likes to be stereotyped. Jesus avoided earthly labels, and we should use caution attaching them to environmental and alternative energy issues. Philippians 2:3–4 teaches: "Do nothing from rivalry or conceit, but in humility count others more significant than yourselves. Let each of you look not only to his own interests, but also to the interests of others." Both political parties could learn from that.

Tough Choices

Actual leadership poses one of the greatest challenges for political leaders. The political process has become reactionary, rather than proactive, which would drive decisions for the good of the country. For example, politicians cannot legislate morality—in the sense that we cannot fully control people's actions, only attempt to prevent bad actions by administering punishment afterwards. But as Christian politicians, they should not condone immorality by inaction, either. If God has given us the mandate to be good stewards of His creation, then we are morally bound to do so. We cannot dismiss it to economic or political rhetoric, nor can we mandate it to force the world to share the earth's resources with equity.

Our country and the world needs true leadership from our government officials—leadership defined by the willingness to look at long-term needs, and not just the political polls of the day or saying what will get them elected or re-elected. A shocking way of stating it is that we need to repent. Set aside the negative connotations of repentance the secular world perceives. By repent, we mean being determined to change for the better. Repent means willingness to surrender to change, to change one's mind. Examining our hearts in comparison to biblical truth needs to happen at all levels, but willingness to repent—to surrender to change called for in God's Word—is especially needed at the leadership level.

In certain respects, we need elected government officials to preserve the status quo. In many regards, we have more comfort of life today than at any time in history. Change is hard to accept sometimes, particularly when asking the public to adopt new alternative energy technologies.

When the White House was first wired for electric lighting, President Benjamin Harrison was in office (1888–1892). He and his wife were both afraid to touch the light switches. It seems strange to us today that the leader of a great nation would fear a light switch, but the fear of change can be overwhelming.

Today, many leaders will not embrace the change to alternative energy systems because of the same fear of the unknown. It is safer politically to stick with the status quo and claim the costs would be too high to change. Change has occurred over the last couple of years as more attention is drawn to the environmental and national security issues swirling around conventional energy sources. Many envision international consensus, referendums, or even laws restricting carbon emissions or requiring renewable energy credits (RECs). In any case, at least the political wheels are churning forward, however slowly, and the political drivers are moving to the front seat.

It's important to recognize God's work in any of our forward advances. God can work miracles with anyone, even government officials and an apathetic electorate. The Bible is full of stories where God accomplishes His will through government and politics, both good and bad governments, including the role government played in crucifying Jesus. If you focus on the horrific role the Roman government and Hebrew political and religious leaders played in crucifying Jesus, you miss the immensely more important act of God's salvation through the resurrection of Jesus. The electorate and the politicians must focus beyond the near-term scenarios for energy and environment. They must act on long-term interests for the good of our children, country, and especially to be in sync with the will of our God.

In the end, we need to remember that Washington D.C.'s population consists of many lawyers who know nothing of energy technology other than what lobbyists tell them. Yet they must lead to decide which roads to build. It's like driving blindfolded with only verbal directions from a self-indulgent three-year-old. Many say the best way for big government to drive energy policy is not to have government pick the technology to promote and protect. When that happens, well-intended protection can hinder the technology. Once protective policies begin, they become a crutch that is hard to throw away. Look at the U.S. farm subsidy program, for example. Intended to support the small family farmer, it more often protects the large corporate farm. If the product needs the protection of big government to succeed, then it doesn't meet the core fundamentals of capitalism.

But in the case of our energy conundrum, big government must lead, as scary as it is. Why? For three reasons: first, energy infrastructure is too big, especially in this world of global economics, to rely solely on market forces to pick the new alternative technology winners. It is too slow and the inertia of the status quo is too compelling for incumbent technology industries to change on their own. Second, environmental policy regulations—especially those related to human-caused greenhouse gases—are tied to energy policy. If you solve problems with energy, you solve the other. Third, the market cannot appropriately value externalized costs; big government must step in with an overarching policy for national security reasons. Reliance on Middle East oil is the best example of that pothole.

Fortunately, a hydrogen economy utilizing fuel cells can solve all these problems. But what about the dangers which result from allowing those lawyers in Washington to pick a winning technology and protect it? The answer resides in the diversity of hydrogen and fuel cells themselves. Hydrogen is available from many sources, and can be stored in several ways. Politicians don't need to pick the winning path of which will be most efficient and cost effective; the market and technology can sort that part of it out, if we have clear direction and know where we want to end up. Likewise, for fuel cells, several different technologies with several different applications are vying for market and technical supremacy. In essence, government wouldn't be settling the details, just bulldozing a rough path to the overall goal.

SMALL GOVERNMENT

We've discussed the strengths and limitations of big government, but many believe that small government at the state and local level can better address the ills of society. We've seen recent steps in the right direction. In many ways, state governments and local communities have taken the lead, providing the services needed locally, like the driver of a city bus.

California has the most aggressive plans in the country, with progressive renewable energy plans slated to provide 20% of the states' energy needs by 2010.[7] Solar thermal power plant developers have queued up to put parabolic mirrored troughs of solar fields in the Mojave Desert, while the Million Solar Home initiative changes the suburban landscape with incentives for solar photovoltaic (PV) panels on residential homes.

In Kansas and throughout the Midwest, wind farms dot the landscape, providing electrical power and alternate sources of income for struggling agricultural communities. Bio-fuels will continue to play a part in rural America and throughout the agricultural world, providing alternate avenues for cash crops. If done in moderation, it can act as a bridging technology to reduce oil dependence while supporting local economies.

Also in Kansas, tragedy has turned into triumph for the small town of Greensburg, which was leveled like a scene from the movie *Twister* by an F5 tornado. Nevertheless, Greensburg is rebuilding "green" from the ground up in an effort to revitalize their community in a way that would have been impossible prior to the storm. Green building standards, such as LEED (Leadership in Energy and Environmental Design), apply to many of the new buildings planned in Greensburg.

Other states have required stricter energy efficiency and green building standards in government buildings statewide. Colorado, Rhode Island, and Virginia require certain levels of LEED certification. (The levels are bronze, silver, gold, and platinum.) Still other states like California, Hawaii, and Oregon have higher building code standards for all new construction.

Virginia committed to cutting greenhouse gas emissions by 30% before 2025,[8] heavily pushing energy conservation and energy efficiency. The Virginia government has also increased research and development funding for alternative energy at state universities.

Although we need a comprehensive energy policy at the national level, state and local governments still need to be heavily involved in energy policy because energy resources are local, especially in the case of renewable energy. For example, you cannot put a tidal energy system in Kansas, and a PV solar installation would not work well in cloudy London. Let the local governments and people decide what makes sense, works best, and utilizes the resources available.

Using Renewable Portfolio Standards (RPS) can help to accomplish better local resource allocation. This regulatory policy, set by government, would encourage or mandate a certain percentage supply of renewable energy from electrical power providers. The electrical power providers can generate the renewable power themselves, or purchase it from others. Depending on how it's defined, an RPS can be mandatory or voluntary, and can cover a wide variety of renewable sources that the local population supports. Voluntary RPS must contain enough carrots (incentives) to encourage utility companies and energy providers to want to follow.

Most people would not know an RPS from a GPS (Global Positioning System), but oddly, they have similar features. Both show where you are, both deal with technology, and both help you get from point A to point B. A GPS knows your physical location; an RPS encourages the use of renewable resources at your location. A GPS gives valuable technical information and electronic maps of your surroundings; an RPS defines an incentive basket of renewable energy technologies available for your surroundings. A GPS helps get you from location A to location B; an RPS helps get you off conventional energy sources and onto renewable energy sources.

A low-requirement RPS should be set nationally to provide a minimal mandate and some consistency from state to state. However, states with more opportunities or local public support in the renewable energy field should set higher goals for themselves. Minnesota, Illinois, and Oregon set 25% RPS goals by 2025. Colorado, Connecticut, Delaware, Hawaii, and New Mexico target 20% RPS compliance by 2020.[9]

The Western Governors' Association, an alliance of nineteen western states, strongly pushes for alternative fuels and clean energy. This encourages cooperation between states and strengthens the regional approach to project development. It's a step between the state and federal level that can be quite helpful.

On a similar note, regional councils (or "power pools") that incorporate several states handle much of the electrical power transmission. Nationally, the grid is tied together and federal guidelines and regulations exist for electrical interconnection to the power grid, but it is interesting how much is decided locally.

Although few consumers know of it, government could encourage the development of locally derived alternative power through *net metering*. Basically, net metering allows customers to "spin their electric meter backwards" and sell electricity back to the power company. Of course, you can only do this if you have a distributed generation source, such as fuel cells, solar panels, or a wind turbine at your house. However, if net metering regulations are written and implemented in a way that provides financial benefits for consumers, distributed generation would become popular much more quickly.

Because most consumers do not know of the possibilities, they aren't asking their politicians for legislation in this area. Of course, the standardization of minimal net metering rules is also necessary for safety of utility workers and for economic fairness to the utility as well as the consumer. Local involvement is necessary because different utilities have their own safety requirements that are familiar, safe, and serve the best interests of their customers. It's like being in an unfamiliar city: the visitor is safe going with McDonalds, but if you want the true flavor of the city, the locals are the people to ask.

From Adam and Eve to Adam Smith:

A laissez-faire, minimal government approach would work fine, if only it were an ideal world. But the perfect world left us when Adam and Eve sinned. If only capitalism or any economic system could handle the intangible factors of godly stewardship . . . but without God, they cannot. Adam Smith, the father of modern economics, did not openly espouse Christianity, but he did acknowledge something more at work than sheer greed. Before Adam Smith wrote his best-known work, *The Wealth of Nations,* he authored *The Theory of Moral Sentiments,* in which he suggests that the "invisible hand" of the economic system is a moral one.[10] Therefore, what drives the markets is not just money, but a moral compass to fulfill obligations of society.

Unfortunately, it is hard to put a price tag on morals or the intangible effects that occur if the moral compass fails. These intangible factors include the true costs to the environment, the quality of health for people downstream of the pollution, national security impacts, and general long-term effects compared to short-term benefits. Without government regulation and restriction of dangers, coupled with the endorsement of safer alternatives, the free market often fails to protect its citizens. One of the key purposes of government is to protect from threats abroad, but also from threats within.

So what is the right balance of political involvement? One man's perceived protection is another man's perceived persecution. Where do freedom and fences balance perfectly?

God has given us a big yard. The Bible tells us that we have perfect freedom within the fence of God's law. Outside the fence of God's law, our perceived freedom feels more like shackles.

Business Drivers

> "Command those who are rich in this present world not to be arrogant nor put their hope in wealth, which is so uncertain, but to put their hope in God, who richly provides us with everything for our enjoyment."
> —1 Timothy 6:17

Wall Street does not live by 1 Timothy 6:17. Wall Street seeks profits, not prophets, unless you count Peter Lynch and Warren Buffet. Does that exclude big business from anything to do with biblical environmentalism and sustainable energy development? Of course not. In fact, we have many concrete examples of how business can impact our energy problems as a direct result of positive consumer demands. For example, the John Deere Corporation, primarily makers of farm equipment, now help farmers harvest the wind, too. In addition to their core business, John Deere founded a new business unit to help their farming customers invest, finance, and develop wind energy projects. The farming customer gets a new revenue stream, John Deere strengthens their already strong reputation, and the electric grid gets clean, renewable power. It's a win-win-win situation.

You might have to read *between* the bottom lines to figure out what moves the market, and maybe it's not quarterly profits. For some companies, it is improving public relations in an increasingly competitive marketplace. Some would say making a positive difference in society—the moral invisible hand first discussed by Adam Smith—still applies today. Although we often ignore it, natural resources and people have value beyond the balance sheet, intangible resources and value given by God. And the resources He gave us include everything we could possibly need. The Bible tells us in Jeremiah 29:11, "'For I know the plans I have for you,' declares the Lord, 'plans to prosper you and not to harm you, plans to give you hope and a future.'" Prosperity, usually affiliated with fiscal success, in broader terms means success in general. Success is making the most with what God gave us fiscally, spiritually, socially, even environmentally.

Jesus' parable of the talents illustrates that concept perfectly. A talent in New Testament times was a coin worth a large amount, about an average person's yearly wages. The story, found in Matthew 25:14–30, tells of three servants whose rich master entrusted his money to them according to their ability and skill. The wisest servant was given five talents; the next, two talents; and the last, one talent. The master went away for a long time, and when he returned, he called for his servants to gather the earnings he expected to receive. The first two servants doubled their money, but the last did not make a positive difference with his talent coin because he hid it. The master took the one talent entrusted to the servant away from him and threw him out because he had done nothing with what had been given him. This may seem harsh to some, because the servant hadn't lost the money, but the fact that he had done nothing productive with that entrusted to him was worse than trying and failing.

The great teaching value of parables lies in all the parallel lessons derived from them. Consider the parable of the talents in terms of earth's resources instead of money. God has entrusted us with His creation, which is infinitely more precious to God than money. Doing nothing with the earth's resources is of great offense, and trying to make a positive difference is encouraged. If the preservation of God's creation comes at the expense of help for the poor, then maybe we should reread Matthew 25.

The other latent teaching in the parable of the talents is that creation and all of God's gifts deserve use for the good and the enrichment of God's kingdom. If we perform good works with God's goodness, we join the celebration. However, if we let that good profit fall to sinful desires, we no longer deserve God's favor. Case in point: if God-given resources produce profit squandered on drugs, pornography, and other detestable things, God does not see that as a positive use of resources.

The best use of resources involves finding a sustainable way to use them for the betterment of everyone, like making hydrogen from wind power. This can make a positive difference with the resources *and* make a profit. The profit may be smaller at first, but the use of the resources is sustainable and therefore, better. In the Master's eyes, that's better than the purely monetary gain which so often falls to the purchase of detestable things.

God does not judge us by how much money we earn, but by what we do with that money and how we got it. The Bible repeatedly tells us that no one can buy their way into heaven, but calls us to use our earthly natural resources, and resources of time and money, to make a positive difference for God's kingdom. Because God's grace redeemed us through our acceptance of Jesus, we are called to do good works. Making a positive financial gain is only part of the story. The positive use of resources and positive application of profits is more important, in God's view. Along these lines, John Wesley preached a famous sermon on the use of money. The basis is stewardship: "Gain all you can . . . save all you can . . . and give all you can."[11]

The deceitful accruement of riches and ill-gotten gains ensnares many people even today. A prominent pharmacist in my community donated a million dollars to help build his congregation's church. Unfortunately, he diluted cancer-fighting drugs at his pharmacy to increase his profits. He tried to do something positive with his ill-gotten gains, but was that really even for the benefit of God's kingdom, or the enhancement of his own image? He now sits in jail, and the church returned his gift.

WHO'S IN CHARGE HERE?

God clearly tells us that government and politics are not the final authority to what we'll do. Proverbs 21:1 states, "The king's heart is in the hand of the Lord; he directs it like a watercourse wherever he pleases."

But it's not just our government leaders who make up the political drivers, it's also the voters. The voting population must hold government accountable, which we have largely failed to do. Voter turnout in the 2000 and 2004 presidential election years only reached a disappointing 51% and 55%, respectively.[12] Years of partisan politics, amongst other things, has led to voter apathy. The founding fathers of our nation would be dismayed at our general lack of interest in the government and the apathy that allows special interest groups to overrule the good of the nation too often.

A CBS/Washington Post poll in April 2008 has shown 81% of the United States population is pessimistic about our future.[13] Health- and wealth-wise, we're better off than any time in history, so maybe the art of pessimism has also peaked. But to the family living paycheck to paycheck, comparisons of life today to medieval peasant life mean little. What matters is the food, or lack thereof, on the table today and in the near future. Concern that the future may suffer because of our choices today remains legitimate; we should be concerned for ourselves and for future generations. Continuing to worry takes a backseat role and won't get the job done.

Consumer Drivers

Yogi Berra, the baseball Hall of Fame legend, attracts as much attention for his sayings, known as "Yogisms," as he did for his baseball prowess. One of his sayings, "I can't think and hit at the same time," speaks volumes of his approach to hitting. The wisdom underlying the obvious humor says that you need not over-think everything; sometimes it's better to just do it—and that's what Yogi successfully did, leading to ten World Series championships and a coveted space in the Baseball Hall of Fame at Cooperstown.

This Yogism has a green corollary for the consumer: I can't conserve and consume at the same time. The wisdom underlying this humor reminds us that we consumers can be very successful at conserving, if we consume the right things and don't consume the bad. For example, if one consumes renewable hydrogen in a fuel cell for power, you're a Hall of Famer. If one consumes oil from the Middle East, you're a utility player on the bench. Both consumers use the same amount of energy, but the hydrogen consumer does it much more successfully, while still conserving limited resources.

However, since inexpensive fuel cells aren't yet readily available, how should the consumer discern the best path to energy bliss? Have you ever felt like you have more information, but are less informed? In today's communication-saturated society of cell phones, 24-hour news channels, e-mail, and the Internet, it's easy to reach information over-load. Now we're also barraged with climate change, claims of growing geopolitical instability in the Middle East, cries to conserve more, re-cycle more, and reuse more or risk being part of a global environmental meltdown. Who needs that on top of all of the day's other activities? As individuals, we all consume, and we all leave a carbon footprint comprised of the amount of energy and stuff we consume. It's hard to know what to believe, and nearly impossible to get our arms around the big picture.*

Consumers are not helpless victims when it comes to the environ-ment or energy matters. We have a vote and we exercise it by how we spend our time and money. Without consumers, business does not exist. It's a two-way street. Consumers need to become wise consumers, making the right decisions although they might not know all the facts. With most consumers blind to energy infrastructure and distribution, knowledgeable decisions are unlikely, though we still have hope that wise decisions can be made without all the facts. In Chapter 5, we

---

* If you have your roots in Christ, you can weather the storm. He is the vine and we are the branches (John 15:5–8). It may not seem like a compilation of writings from thousands of years ago could address modern day problems, but the Bible does it for us. Its truth is eternal and relevant to all times, places, and people.

discussed how chaos theory might preclude us from ever having 100% certain knowledge of the causes of climate change. This explains why we cannot wait for knowledge to lead us out of the energy conundrum; we will have to hope and pray for wisdom.

However, when energy professionals don't know the path forward, confidence shrinks. It doesn't help that their past predictions have proven to be as accurate as a Midwest weather forecast. We can become paralyzed by fear, but just as Peter walked on the water to join Jesus on the stormy waves, we will need a large measure of courage and faith. Remember, Jesus calls us to action, and we have to get out of our comfort zone if we want to walk on water. With Jesus' help, we can conquer our fear if we stay focused on Him instead of all the negative things surrounding us. God never calls us to be anything other what He made us to be; however, we often don't realize all that we're made of until we're tested.

Individual consumers often feel like little David facing giant corporate and governmental Goliaths. But consumers need to remember that David won! Moreover, the increasing numbers of consumers concerned about energy issues can have the same type of unlikely victory. As consumers, we can make a positive difference if we:

- Purchase and use energy efficient products. This can be as simple as adding insulation, sealing air leaks, using your own cloth shopping bags (instead of paper or plastic), or buying toilet tissue made with recycled content. It signifies our willingness to change our consumption patterns until the adoption of better energy supply options.
- Educate ourselves on alternative and renewable energy sources. We don't have to know it all, but some knowledge will help us make wiser decisions and embrace better energy supply options when they come along.
- Encourage our governmental and business leaders to educate themselves on the benefits of alternative energy and the responsibility of energy efficiency. We can guide them to wise decisions by voting with money and our political votes.

- Consider purchasing solar panels, solar hot water heating systems, or other clean energy producing or energy saving equipment for our homes, as personal finances allow.

As individuals, we sometimes feel as if we are trying to boil the ocean. What do you think the scene looked like when young David went out to battle the armored giant Goliath to the death? Pretend you don't know the outcome of the story. It wouldn't look good for David or the Israelites, would it? But thankfully, logic and reason do not always win. Conventional wisdom would say to send the biggest warrior the Israelites had out to face Goliath, but biblical wisdom said to send David. The biggest challenge for consumers and individuals is to look beyond the conventional energy and products of today and demand better, unconventional alternatives. In my opinion, fuel cells will soon fulfill the call for an unconventional wisdom solution; they are the David on the horizon.

SCHOOL BUS DRIVERS—CHRISTIAN DRIVERS

When it comes to energy and the environment, too many Christians just go along for the ride, with Thelma and Louise at the wheel. Many Christians understand the urgency, but more need to join their ranks and take action. If we truly want to lead others on the road to energy and environmental stewardship, Christians must get behind the wheel of the school bus, a job with great responsibility, but little appreciation. God blessed the United States with many freedoms: economic, spiritual, and social. We have been given much, and much will be expected of us, and not just from the wealthy. Luke 12:48 quoted at the start of the chapter refers to much more than that. We all have spiritual gifts (1 Cor. 12:4–11) and we must use them. If Christians use their spiritual gifts, they will pursue whatever means they have to be more like Christ. We cannot sustain His church if we destroy the world around us. To honor God and to use our spiritual gifts, I'm convinced we should embrace wise solutions to energy and environmental problems, which can only assist us in leading people to Jesus. Good stewardship in energy usage and environmental care transcends our station in life. Christians *must* use their spiritual gifts to glorify the kingdom of God on heaven and earth if we want to follow the Bible.* Good stewardship demonstrates our love for God and provides a better life for our fellow man, future generations, and ourselves.

In the past, the Christian political movement has concentrated its fight on two core, traditional issues: abortion and gay marriage. But for many Christians today, the debate has grown to include key issues such as social justice, compassion for the poor, and ecological stewardship. Evangelical leader and political activist Reverend Richard Cizik points this out, referencing a poll of U.S. evangelicals. In the survey, 52% of evangelicals believed the unborn sanctity of human life was a priority, but 48% declared that a clean environment was also a priority.[14] Some see this move as unnecessarily deviating from the core focus; others see it as a move to the broader biblical call we should have followed all along.

In this world of sin, we fail to be perfectly holy; however, we should always strive to follow Scripture as closely as we can. James 1:22 says, "Do not merely listen to the word, and so deceive yourselves. Do what it says." In that regard, we cannot help but feel compelled to act on social justice issues, care for the poor, and environmental stewardship. Deuteronomy 15:11 says, "There will always be poor people in the land. Therefore I command you to be openhanded toward the poor and needy in your land."

Politics provides one way, but not the only way, to advance the rediscovered frontier of the Christian care agenda, especially for the environment. It is a shame that more Christians on both sides of the political spectrum are not working together on growing stewardship of God's creation. Christians should examine the foundation of their faith, make sure they're grounded in their faith, their branches strongly attached to the vine of Jesus. Then they will be prepared to go out into the world—whether among secular or new-age environmentalists, or to big business corporations, or the bowels of big government—and make a difference for the Lord. Not for personal gratification—although the

---

* The earth was cursed by our sin (Gen. 3:16–19). Jesus came and redeemed us and all of creation (John 3:16), but all peoples and the earth are still subject to sin until Christ comes again (Rom. 8:18–25). Until then, we continue to battle sin. Although we live in a sinful world, we are still required to care for creation as God has commanded.

Lord might reward you in that way—but for the wisdom of following God's command in Genesis 1:28 as a way of bringing your daily life in line with God's will.

In Christianity, everything points to the cross, where Jesus Christ died for our sins and rose again, and God restored not only us, but his whole creation by Christ's resurrection. One of the most popular verses in Scripture, John 3:16, confirms it: "For God so loved the world, he gave His one and only Son, that whoever believes in Him shall not perish but have eternal life."

The Greek word for "world" in the early translations of John 3:16 is *cosmos,* defined as the whole universe. In Christian terms, this means God's whole creation, of which we are the pinnacle. God renewed His whole creation through Jesus. We anxiously await the return of Christ and the full restoration of God's creation. Of course, we should be cautious not to take love of God's creation too far (Rom. 1:25). Nevertheless, our call to be good stewards of God's creation—to "work it and care for it" as given to Adam in Genesis 2:15—is renewed through Christ.

Knowledge equals power, the old axiom claims. But wisdom brings humility to the powerful. What is wisdom? Or more importantly, what is biblical wisdom and how should Christians apply it? Specifically, how does biblical wisdom apply to energy and the environment, so that Christians can see their need to lead?

First, let's start with plain old wisdom, looking at the Wikipedia definition:

> "Wisdom is the ability, developed through experience, insight and reflection, to discern truth and exercise good judgment. Wisdom is sometimes conceptualized as an especially well-developed form of common sense. Most psychologists regard wisdom as distinct from the cognitive abilities measured by standardized intelligence tests. Wisdom is often considered to be a trait that can be developed by experience, but not taught. When applied to practical matters, the term wisdom is synonymous with prudence. Some see wisdom as a quality that even a child, otherwise immature, may possess independent of experience or complete knowledge.

> Contemporary culture limits the importance of wisdom and intuition, particularly in American culture; Wisdom/Intuition is a right-brain activity, while logic is a left-brain activity. . . . Some define wisdom in a utilitarian sense, as foreseeing consequences and acting to maximize the long-term common good."

The description of wisdom above contains many good points. However, biblical wisdom encompasses even more. The biblical book of Proverbs provides an excellent place to look for God's wisdom, much of it given through King Solomon. The beginning of Proverbs explains the goal of its writings, which many believe are especially intended for young adults. Proverbs 1:2 says, "For attaining wisdom and discipline; for understanding the words of insight; for acquiring a disciplined and prudent life, doing what is right and just and fair . . ."

The New International Version Study Bible relates to us that skill is the basic idea behind wisdom. The term is used in the Old Testament to describe the abilities of tradesmen and professionals of all types, from garment makers to goldsmiths. The book of Proverbs uses the word "wisdom" to speak of the skill of living life in a way that honors God. A wise person has the ability to adapt his or her life in a way that honors God. A wise person has the ability to adapt his or her life to God's pattern. A foundation of reverence for and worship of God distinguishes biblical wisdom from the so-called wisdom of the other ancient nations. Because God is the source of wisdom, reverence for God is the controlling principle for applying these wise observations about the way life works. Wisdom is the cornerstone of Proverbs, and we can all learn to incorporate godly wisdom in our jobs, schoolwork, and everyday skills, including making the right energy and environmental choices.

Why, then, do so many Christians seem to avoid God's mandate to govern wisely over the earth and its resources? Since God tells us in Genesis 1:28 to subdue (govern, control, to use but not abuse) the earth, and God teaches us in Proverbs that our daily life decisions should apply godly wisdom and skill, we should obey God's command. We sometimes seem to pick and choose which of God's commands we want to follow, and too often ignore the inconvenient purposes He has in store for us, including our biblical responsibility to the care of His creation.

Jesus taught in Luke 12 about being watchful:

> The Lord answered, "Who then is the faithful and wise manager, whom the master puts in charge of his servants to give them their food allowance at the proper time? It will be good for that servant whom the master finds doing so when he returns. I tell you the truth, he will put him in charge of all his possessions. But suppose the servant says to himself, 'My master is taking a long time in coming,' and he then begins to beat the menservants and maidservants and to eat and drink and get drunk. The master of that servant will come on a day when he does not expect him and in an hour he is not aware of. He will cut him to pieces and assign him a place with the unbelievers.
>
> "That servant who knows his master's will and does not get ready or does not do what his master wants will be beaten with many blows. But the one who does not know and does things deserving punishment will be beaten with few blows. From everyone who has been given much, much more will be demanded: and from the one who has been entrusted much, much more will be asked."
>
> —Luke 12:42–48

This parable Jesus taught regarding the end of times and the need to be watchful ties to our energy and environment situation, too. We need to be watchful of how we use energy and correspondingly treat the environment. God will judge us on how we have lived and treated His creation. As Christians, he expects more from us than the rest of the world. We need to lead!

However, God is patient, and there is no reason to ignore His commands any longer. "The earth is the Lord's and everything in it, the world and all who live in it" (Ps. 24:1). Christians need to get behind the wheel and drive this cause wisely down the road God intends. And Christians shouldn't do it with tinted windows; we should allow the world to see, for God's glory, not our own! When non-Christians see Christian inaction on energy and environmental issues that affect everyone on the planet, we project an image that our God does not care about creation or the 6.7 billion earthly inhabitants, and convey an image that He is not powerful enough to do anything about it.

Unfortunately, in modern Western society, the role of wisdom has diminished in preference to logic. Logic works fine for a computer, but logic, like knowledge, will never subvert wisdom in prominence. The world doesn't suffer a lack of logic, but a lack of wisdom, evident from the many messes in which humanity is embroiled.

Today, the knowledge and, more importantly, the wisdom of the effects man and nature have on each other make it imperative that we seek God's will both individually and collectively. God has given us the role of steward, not owner, of the earth. A good steward has to deal with consequences arising from his own actions, and also with things outside of his control. The environment we live in is no different. When a flood comes down a valley, it matters not if the dam failed because of faulty engineering or a natural overabundance of rainfall. You must still get to higher ground. As Christians, we should always seek that higher ground.

If we first live as a follower of Christ, wisdom will follow. We will develop the skill to live our lives in a way that honors God. Whether as a consumer, a businessperson, or even a politician, we have the skills to drive energy, environment, and economics in a way that honors God. The market-based society in which we live gives us the opportunity to drive home solutions from both the demand side and the supply side of the market. We honor God by simple actions on the demand side of the equation, such as reduce, reuse, and recycle, but that is not enough.

We need to honor God by working the supply side of the equation, too. That means supporting and switching to alternatives that provide clean energy and are environmentally sensitive, even if they are not economically cheaper at first. The rewards will come in intangibles: cleaner water, cleaner environment, fewer natural resource conflicts, sustainable and strong economies, energy security, all of which honor God if we learn to supply clean energy, best exemplified by the properties of hydrogen and fuel cells.

WOMEN DRIVERS

> "Charm is deceptive, and beauty is fleeting; but a woman who fears the LORD is to be praised."
>
> —Proverbs 31:30

Finally, I'd like to focus on the role of women drivers. Frankly, women as a whole have been denigrated in their role as drivers. However, per historical Bible events, the reality is that women do lead—even if society restrains them. Men dominated society in all nations during Bible times, yet many strong and influential women are mentioned prominently in the Bible. Compared to other civilizations of the day, that's extremely unusual. What an opportunity for women to contribute to the world today through promoting alternative energy—unexpected maybe, but completely within character. Of course, women can do this through politics and leading big companies too, but I think they will shine on the personal level, just like Christ did.

Women make up more than 50% of the world's population, equating to at least 3.25 billion people. In Western society, women have more freedom socially and economically, and therefore play a huge part in our energy choice future. However, one could argue that women, collectively, have a low understanding of the supply side of energy, and could be even worse than men at taking electricity, oil, and gasoline for granted. Arguably, the vast majority of women care more about personal relationships, family, and matters of the heart than energy supply and demand. This generalization is probably correct, and those tendencies could cause women to champion alternative energy issues—through personal connections—by winning hearts and minds to the need for change.

One of the strongest cases by far for the need to change our energy and consumer lifestyle, is the world we will leave our children. As mothers become acutely aware that their children will likely go to war to secure dwindling resources such as oil and unpolluted water, they will, in turn, lead the charge for alternatives. As women sense the social injustice and suffering caused by our addiction to oil, they will search out alternatives with great conviction and passion. Fathers will feel the pain too, but I believe mothers feel, and love, on a whole different level.

These wars over resources may not be blatantly apparent now, but they already exist. Both of the recent Iraq wars latently drip with oil as the subplot. As time moves on, world population grows, and resources such as oil wane, resource wars will grow and intensify. Future generations—our sons, daughters, and grandchildren—will have to deal with this situation. The author of *The Limits of Power*, Andrew Bacevich, reflects these sentiments regarding the American lifestyle and demand for resources.

> "The collective capacity of our domestic political economy to satisfy those appetites has not kept pace with demand. As a result, sustaining our pursuit of life, liberty, and happiness at home requires increasingly that Americans look beyond our borders. Whether the issue at hand is oil, credit, or the availability of cheap consumer goods, we expect the world to accommodate the American way of life."[15]

If mothers realize they can switch to alternative sustainable energy solutions, ones that remove the drive for senseless resource wars over oil, they will protect their children swiftly by doing so. When preserving our consumer lifestyle necessitates sending our children to war to ensure energy and economic "freedom," we need to change. An ounce of prevention is worth a pound of cure.

One reason women make great drivers for change is that many believe women are more compassionate than men, and more attuned to social justice. A recent Pew Research Center survey shows that 80% of Americans think women are more compassionate.[16] Obviously this will show in women's attitudes to the future of their children, but other results will flow from their compassion.

For instance, environmental degradation affects the health and living conditions of the poor, the young, and the old more than it does the rich and affluent, both locally and throughout the world. The rich can move away from pollution sources and flood-prone land, if they choose. The poor do not have that luxury. Women connect compassionately to their plight when they become aware of it.

As a group that composes more than half of the world's population, women represent a viable source for change. As more women discover the alternatives possible with clean energy supplies, and the familial, social and environmental benefits they bring, I believe women will embrace the wisdom of green stewardship through alternative energy. For example, the women I know are arguably more idealistic and less attuned to the mechanics of energy infrastructure than the men. But they shouldn't be dismissed as being unable to discern or unreasonable to the logistics of energy supply; they just judge what is valuable to them in a different way. To them, family, personal connections, and compassion are much more important. When the benefits of sustainable alternative energy strike home, I believe women will recognize the value and react sooner than most men. I admire that quality, and I believe that's how Jesus would want us to react.

Women have an ever-increasing influence in Western society, and their idealism is motivational, especially in family and grassroots efforts. As paraphrased from the movie, *My Big Fat Greek Wedding,* "If men are the head of the family, then women are the neck. The neck can turn the head any way it wants." A groundswell of support by women can make a big change in the products, services and decisions offered in the commercial world. Women can have influence, even if it appears too many facts are unknown to drive a positive change.

To humorously illustrate this case, let me tell you a story of the one and only time I bought a purse for my wife. It was a surprise Mother's Day present from my son, a toddler at the time. My wife needed a larger purse for all the mommy stuff, so I started searching for dimensions, cubic volume, and weight capacity—you know, all the logical information a person would need to make the best decision. (Remember, I'm an engineer.) Guess what, not one purse gave any of that information on the tag of the purse or inside of it. I thought, "How impractical! How do women buy purses without vital information like this?" The only information the tags gave was the designer name, and maybe the color. I wondered, "What good is that? I can see the color." But women know what they like, and what they need. Most end up with a good purse that works well with the majority of their shoes. By the way, I ended up purchasing a dark blue one with brown piping. It was a success, but I vowed, never again.

The point of this story: you don't have to know all the facts to make a good decision. Many women do it every day, even when deciding on a purse. In regards to the climate debate, why should we require scientific certainty before deciding to be good stewards of God's creation? We shouldn't need a "climate hammer" to hit us over the head. Common sense and asking, "Is this being a good steward of God's creation?" will lead to wise, clean, and sustainable choices.

Proverbs 14:1 states, "The wise woman builds her house, but with her own hands the foolish one tears hers down." Notice that the wise woman works within her family, at the grass roots level, and makes a strong foundation. This is how Jesus worked also, laying the foundation of Christianity with 12 disciples. Today, Christ has more than 2.1 billion followers. Although not documented in the Bible, many believe Christ traveled fewer than 100 miles in radius from Jerusalem during the three years of His active ministry. Jesus worked with individuals, and even closer with his inner circle of disciples: especially Peter, James, and John. If Jesus worked hard at building those personal relationships, then we should also.

Women are exceptional at building personal relationships. Women could excel by using these skills in building connections to make better choices in energy, and provide better care for the environment. Compassion at work, hearts and minds changed forever—no climate hammer required.

Good Drivers Wanted

Many groups have a stake in improving energy supply and use. Some of the most influential drivers, such as the government or big business, may hold the keys to the future, or at least they think they do. In the end, the difference between the value of earthly knowledge and the value of biblical wisdom will determine where we go. That's why it's so important for all of our drivers to understand the limits of worldly knowledge.

Wisdom eats knowledge for breakfast every day of the week. You can have all the knowledge in the world, but if you don't use it wisely, you fail. It makes no difference how smart you are; you can end up making colossal blunders. Think of Ken Lay and Jeff Skilling of Enron:

smart guys, but what a colossal mess. You can also make the case that Bill Clinton possessed more book smarts than any president in recent history, but that didn't stop him from making a colossal mistake with Miss Lewinsky. Bernie Madoff had ample knowledge, money, power, and respect, but that provided no lid for the pot of greed he cooked up with his Ponzi scheme. You can have no knowledge of a situation and make a wise choice. Look at the story of King Solomon and the two mothers fighting over a baby.

The story can be found in the biblical book of 1 Kings 3:16–27. Each mother had a baby, and one mother rolled over and accidentally killed her baby in the middle of the night. She then switched babies with the other sleeping mother. Of course, there was a huge argument the next morning, and the squabbling mothers were brought before wise King Solomon. In what appeared to be a cruel solution, King Solomon ordered one of his guards to cut the baby in two and give half to each woman. The mother whose baby had died agreed. The live baby's mother begged the king not to follow through with his decree and give the baby to the other woman. King Solomon gave the baby to the woman who begged for the baby's life, recognizing that the true mother would rather see her child given to another than destroyed.

Although the wise king didn't originally have knowledge of the real mother, he made the right and wise choice. Solomon did not have the scientific knowledge of a blood test or a DNA test, as this event happened approximately 3000 years ago, but he possessed wisdom. Solomon recognized at an early age the superiority of wisdom over knowledge, and sought out the gift of wisdom through prayer.

Likewise, Christians should also seek wisdom in everything, including how to use Earth's resources. We don't need complete knowledge; we only require knowledge of God's commands and how to apply them. Remember the old axiom, "A little knowledge is a dangerous thing." The first instance of this axiom goes back to the book of Genesis, chapter 3, verses 4–5. Man fell into sin when Adam and Eve ate from the Tree of *Knowledge* of Good and Evil. The serpent told Eve she would have knowledge, like God, of good and evil. Of course, the first thing Adam and Eve realized after committing the first sin was that they were naked. That wasn't the knowledge they expected. They may have gained worldly knowledge, but definitely not the God-like knowledge and wisdom they anticipated. Instead, they gained knowledge of God's wrath. What a high price to pay for a piece of fruit.

CHAPTER 10

CUT TO THE CHASE: POSSIBLE SCENARIOS FOR A PATH FORWARD

> Now to Him who is able to do immeasurably more than all we ask or imagine, according to His power that is at work within us.
>
> —Ephesians 3:20

The energy challenge that lies ahead requires patience and perseverance, not fear, but hope, not perpetuating the past, but forming the future. With God, everything is possible, if we sincerely ask for His guidance and obey His commands.

The Lord gives us a variety of tools to use, and we should use them all. But as the saying goes, when you have a hammer in hand, every problem begins to looks like a nail. We need to stop picking up the hammer for every problem—or the drill, in our case. Instead of "drill, baby, drill," for oil, put the drill down, and keep new oil drilling to a minimum while preserving the precious oil remaining for future generations.

Oil is too valuable as a chemical feedstock to other products to burn it up as transportation and heating fuel when alternatives such as hydrogen could replace oil as a transportation energy carrier. However, our energy conundrum extends beyond oil. In order to transition to a sustainable energy future, we need to diversify all energy resources.

201

Energy is an enabler. On the dark side, bountiful energy resources enable and empower Islamic radicals in the oil-rich Middle East. Closer to home, energy use has become the ignored backbone of our economy. Energy consumption distinguishes poor nations from rich nations. America cannot sustain the present consumption mix of energy sources indefinitely. Nor can America use incumbent technologies to bring developing nations out of abject poverty and up to the same level of energy usage and prosperity that we enjoy; at least, not without collapsing the world economy and destroying the environment in the process. Many estimate that if the 6.2 billion people living outside the United States (95% of world population) consumed energy and products like we do, we would need the equivalent of three to five planet earths to supply the resources.[1]

Since acquiring and using energy is such an important part of our existence on earth, albeit underappreciated, we need to examine our way of life. We should ask ourselves these questions: "Can we sustain these practices for our lifetime and our children's lifetime? Or will our choices push our environment over the brink, harm the poor, and squander irreplaceable energy supplies?"

However, the United States historically lacks vision and planning when it comes to long-haul energy policy and sustainability. James R. Schlesinger, the first Secretary of Energy, had this to say when speaking about America's approach to energy in 1977: "We have only two modes—complacency and panic."[2]

Today, our approach to energy resources is not much different. If we listen to big business and the perpetuators of the status quo, complacency will kill us. We experienced the painful results of complacency already when we ignored the risky business deals perpetrated by the credit markets, investment banks, and sub-prime mortgage lenders. We cannot afford to leverage our future with unsustainable energy sources while automakers, oil companies, big business, and government lull us into complacency again.

However, if we listen to peak oil theorists or hardcore environmentalists, then anxiety and panic will ensue. They'll sow the seeds of fear, fertilize them with a healthy dose of skepticism, and soak them with hopelessness. Their crop yields no answers; at least, no answers without a lot of pain.

Neither should we start throwing money at the problem, with one big government program after another. In short, we need a plan based on a strong, sustainable vision and not influenced by complacency or panic.

Vision is exponentially more important than goals. Jesus inspired His disciples and others by His vision, focused on God's truth. Jesus never set goals for His disciples, like save 100,000 people within ten years. Christ never gave "saves" quotas to his disciples like businesses give sales quotas to salespersons. He set a vision without end—"Therefore go and make disciples of all nations" (Matt. 28:19)—not limited by area, time, or quota. Christ focused on the quality of love and faith in personal relationships and not, "We need 100 new converts from Jerusalem by next Tuesday or we won't make our market projections." Jesus never made goal statements like that, but he gave his disciples a timeless vision of how to live according to the will of God the Father.

It's not enough to have vision; you need the right vision, also, like Christ provides for us. The leaders of Enron had vision, but it was the wrong vision. Their vision drove them to pillage and plunder for the purpose of personal power and wealth, and in that respect, they succeeded.

What vision does government and industry have for energy? Is it to pillage and plunder, or is it beneficial for the future? We need to wake up to the limitations of fossil fuels and inspire a sustainable vision with regard to energy. We will need to use transitional resources and sustainable resources to do so. Noted author and columnist Thomas Friedman stated on the David Letterman show, "A vision without resources is a hallucination."[3] The leaders of Enron definitely hallucinated when they leveraged the company way beyond its resources, just like mortgage lenders in the 2008 credit meltdown hallucinated that their house of cards would not collapse. We hallucinate if we believe the status quo of unsustainable energy sources is acceptable for our long-term plans. A true vision focuses on resources sustainable for the long-term.

A Vision with Resources

A world without energy is like a pizza without a crust: a mess without any support. It is painfully inconvenient to focus on energy when so many other topics—like the economy, the environment, or politics—seem more important, but we must realize the crust of the pizza is energy. The other favorite topics, or "toppings" to the world pizza, are debatable, but without the energy crust, we have nothing but a mess.

With so much uncertainty, information, and misinformation out there on the energy front, we need to cut to the chase on ways to proceed "building the roads and training the drivers" for the future energy infrastructure. The world cannot afford to make the same mistakes it made with the overleveraged credit market crash in 2008. We're not there yet, but if we overextend the supply capabilities of the fossil-fuel-based economy with no safety net of alternatives, we will truly see a worldwide depression unlike any other.

Matthew R. Simmons presented a talk at the Association for the Study of Peak Oil & Gas (ASPO-USA) conference on September 22, 2008, entitled *Grappling with Energy "Risk."* Going over lessons learned from the financial crisis, many of the failures are directly applicable to the energy industry: namely, risk is real; leverage can be dangerous; audited financial reports do not always represent genuine numbers; and the greater the risk, the faster big systems can fail. Simmons highlights that the recent financial crisis is fortunately all about paper assets and liabilities. Simmons claims that few people grasp the ominous parallels of the financial crisis to peak oil and gas, and how savage, swift, and devastating a post-peak oil world would be.[4] Man created the current financial mess out of paper, falsehoods, and greed, but a peak oil crisis, based on actual depletion of physical resources, is a thousand times more dangerous.

While ASPO does great work to educate people of the risks of peak oil, they lack the sense of hope that comes from knowing God is in control. Ephesians 3:20 assures us that through the strength of God, more is possible than we ever asked or imagined. God may not give us everything we ask for, but rest assured, if we sincerely ask and pray, God will give us everything we need. Christians should pray for a smooth transition to a sustainable hydrogen economy, however God's plan unfolds, praying specifically for:

• Loving God, loving our neighbor—The law of love

- Energy diversity
- Distributed energy
- Building deliberately toward a hydrogen and fuel cell economy.

Love God, Love Our Neighbor: The Law of Love

The secular world ignores *love* as the most important condition necessary to live a successful and purposeful life, and frankly, many Christians need reminding also. Many other priorities forsake love as the number one priority, such as: money, jobs, security, physical pleasure, even happiness, family, and friends. To paraphrase the Apostle Paul from 1 Corinthians 13, we could have everything, even the faith to move mountains, but without love, we are nothing.

When talking about love in this context, we mean the Greek root word *agape* type of love. It is very different than the "I love pizza," or "I love her body" types of love. Agape is the highest love, a divine love, a love fueled by true need and not desire. Agape love is not rooted in feelings or desire, but in dependence and obedience to God.[5] It is how God loves us, and how Christians should strive to love others. It is the only type of love strong enough, complete enough, to change us into people capable of loving even a fraction of how deeply God loves us. God's love is so deep, that even though we were still sinners (like filthy rags in the eyes of God), He sent His Son to die for us (Rom. 5:8).

God certainly has the capability to love the unlovable. But what about showing love to those evil terrorists who forsake Jesus as the Messiah and afflict pain on our nation from their oil-rich lands in the Middle East? How can we possibly love those we're at war with, and those who kill and oppress innocent people? Love does not mean we must agree and give in. We can show love for people without condoning their beliefs and actions. We don't need to agree with and submit to them to show love. Terrorism is evil—period. Showing love doesn't mean allowing evil. In the end, loving our enemies means we do what is best for them, respecting others as human beings, regardless of what they deserve.

Regarding the terrorists, what's best for them is stopping them from propagating the message of hate. People who live in countries harboring terrorists typically don't condone radical Islamic terrorism. People trying to live their lives in peace deserve freedom from fundamentalist regimes and militia groups who warp the truth of God. Showing love in the Middle East starts with the internal victims of terrorism, but what is the best way to help them? We need to stop empowering radical Islam through oil (today) and natural gas (tomorrow). We can disable terrorists by finding and using alternative fuel.

Terrorist ideology sucks financial support like a mosquito from petroleum dollars. We cannot rely on hoping to identify and kill every terrorist sympathizer in the world and take their oil, nor can we afford to enrich radical Islam further by purchasing their country's oil. What does propagating dependence on oil solve, even if we temporarily get it from outside of the Middle East? It only breeds another generation of hostility. However, we cannot lay down our arms and give Osama a hug. That is not the answer. We must remove the dependence that enables them and ensnares us.

The apostle Paul writes in Romans 12:17–21,

> Do not repay anyone evil for evil. Be careful to do what is right in the eyes of everybody. If it is possible, as far as it depends on you, live at peace with everyone. Do not take revenge, my friends, but leave room for God's wrath, for it is written: "It is mine to avenge, I will repay," says the Lord. On the contrary: "If your enemy is hungry, feed him; if he is thirsty, give him something to drink. In doing this, you will heap burning coals on his head." Do not be overcome by evil, but overcome evil with good.

The Bible does not tell us that our leaders or military are repaying evil with evil in Iraq or the Middle East. It by no means implies we should rebel against our government or abandon our troops who are fighting in the Middle East. Love does not mean we should lay down our arms and passively let our enemies enslave us unjustly. God tells us in Scripture that judgment belongs to Him. No one, not even Christians, can judge the hearts of others, but we can judge their actions and react accordingly. We know the terrorists are evil by their actions. We must continue to fight the wars we're currently in, as long our roots are in righteousness. In the future, though, removing the root causes and enablers of the hatred would better fulfill God's desire for us to love our neighbors. In the Middle East, removing our deathly dependence on oil would help both Muslims and Christians. In the meantime, we should support our troops *and* our government, even if we don't love everything they do.

Paul writes in Romans 13:1–2 regarding submission to government authority: "Everyone must submit himself to the governing authorities, for there is no authority except that which God has established. The authorities that exist have been established by God. Consequently, he who rebels against the authority is rebelling against what God has instituted, and those who do so will bring judgment on themselves."

From Scripture, we clearly see the role of love is much bigger in life, in government, in war, even in loving our enemies, than the world's logic would apply. The secular world operates differently, in accordance with earthly wisdom and worldly logic. Ultimately, too many Christians go along with it, even when it comes to economic and natural resource greed. Economic Darwinism, the "survival of the fittest" mode which permeates societies around the globe, buries the human need to give and receive love near the bottom of the list of life's priorities. "Going deep" from the economic viewpoint is deep offshore oil drilling, not seeking deep love, an agape-type love. Love is seen by many as a want, not a need. However, Jesus taught us this agape love should be the foundation of everything in our lives.

How the greatest commandment or "the law of love" (Matt. 22:37–39 and Mark 12:29–31) applies to energy and environment is hopefully clear. Alternative energy helps us follow and build a road in line with God's will and purpose for His creation. It provides us with an opportunity to *show* our love and respect for God and the world He loves, and to show our love for our neighbor. By curbing wastefulness, increasing supplies of clean energy, and reducing consumption of dirty energy sources, we show love for God and others. Notice that we're not earning God's love, respect, favor, or attention, nor likewise seeking those things from our neighbor; we strive to do them *out* of love, not *for* love.

We don't show love and respect for God to earn salvation or God's favor. God does not need us to make His will come true. We need Him; He doesn't need us. But He loves us and wants us to love, to have faith, to hope, and to follow His commands. God gives us free will, but our pursuit of happiness and love peaks when we follow God's commands, and bottoms out when we fight them.

The second part of the law of love shows the relationship we should have with our neighbor. Our neighbor, as Jesus taught in the parable of the Good Samaritan (Luke 10:25–37), includes everyone who needs mercy. Everyone needs mercy, including our enemies.* We should act as a neighbor not just towards fellow Christians, but to Muslims, Hindus, atheists, blacks, whites, Asians, republicans, democrats, the young, and the old. Our neighbors are people of every race, creed, and color.

Future generations should also be included as future neighbors. What kind of world do we want to leave for our children and the rest of society? We need to drop the philosophy of "whoever dies with the most toys, wins." It's true that you can't take it with you, as the saying goes, but that doesn't mean we should ruin it for everyone else, either. The children and grandchildren we leave behind deserve more than a monetary inheritance. In fact, a healthy planet and a peaceful world would provide a much better gift of love.

If the 2008 crash taught us anything, it should be the fickleness of paper fortunes. Over-inflated values of homes and businesses, overleveraged banks and insurance companies, grossly overpaid company executives, and devious market short-sellers who went too far manipulating the markets breed greed, not love and stewardship. Teaching our children the greatest commandment, and leaving them a planet worth living on with the sustainable resources to carry on, is much more valuable.

It's not enough to agree with the law of love; we must act on it. Action on alternative energy is not a requirement for earning our way to heaven, but we are required to do good deeds. Scripture tells us in James 2:26, "As the body without the spirit is dead, so faith without deeds is dead." For Christians, our actions should speak louder than our words that we value God's creation and love our neighbor.

The law of love applies to how Christians should think and act regarding energy and environment, providing the guiding principle to a successful conversion to better alternative energy sources. How to make it happen when even the energy experts don't agree presents the

---

* Luke 6:27–28 states "But I tell you who hear me: Love your enemies, do good to those who hate you, bless those who curse you, pray for those who mistreat you."

next challenge, not to mention the economic cost and political will required. That's where the next focus should be, now that we have our hearts and minds in the right place.

Energy Diversity

Many people understand the demand side/consumer side of the energy economics equation. They comprehend reduce, reuse, recycle, and quite a bit about energy efficiency and conservation. They struggle a lot more with the supply side/production of energy side of the equation. People want cheap, reliable, clean energy sources. If they have to settle, they go for cheap, reliable, and cheap. However, cheap does not equal best in the end, when supplies dwindle and environmental degradation occurs. Today, about 86% of our total energy sources in the United States comes from fossil fuels; namely oil, natural gas, and coal.[6] Overall, fossil fuels are cheap and reliable, but not particularly clean; hence, people react just as you would expect. Money talks.

However, if the world is really at or near the peak oil plateau, like some industry professionals believe, diversity is absolutely necessary. The supply and environmental issues of fossil fuels are very real. We cannot sustain 86% of our energy supply coming from fossil fuels, coupled with a growing total usage of energy. Something has to give, and we cannot replace one fossil fuel (oil) with other fossil fuels (natural gas and coal). That just exchanges one set of problems for another.

However, the size of our energy infrastructure mandates that fossil fuels will continue to play a large role for years to come, much to the chagrin of radical environmentalists. In addition, many alternative and renewable technologies remain unproven on the commercial scale. Is alternative energy ready for the big time? Is it reliable enough? Is it durable enough? If technically viable, how do we make the transition quickly enough to help the environment, and gently enough to protect the economy?

Renewable energy generation of hydrogen for use in fuel cells could accomplish all this, and we could make noticeable inroads by the year 2019. The National Renewable Energy Laboratory (NREL) prepared a Technical Report in February 2007 detailing this, called *Potential for Hydrogen Production from Key Renewable Resources in the United*

States.[7] The figure below is a redrawn version of "Figure 19" from the report, showing state by state the potential from renewable hydrogen production. When compared to gasoline consumption in the state, it is amazing how much potential renewable hydrogen has to offer.

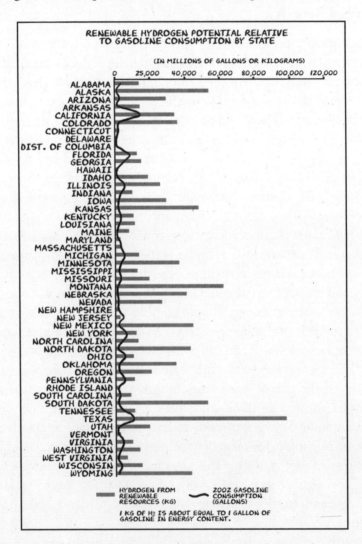

Figure 10.1—Renewable Hydrogen Potential.

(Courtesy of National Renewable Energy Laboratory—*Potential for Hydrogen Production from Key Renewable Resources in the United States*, A. Milbrandt and M. Mann, NREL/TP-640-41134, February 2007, p. 20, Figure 19, http://www.nrel. gov/docs/fy07osti/41134.pdf)

CUT TO THE CHASE: POSSIBLE SCENARIOS
FOR A PATH FORWARD

Not only is it scientifically possible, renewable hydrogen production measures up to other important benchmarks, most significantly biblical stewardship. But it also measures up well against a secular yardstick: small environmental footprint, increased national security, available small-scale to large-scale generation and diverse methods of generation. I predict that renewable hydrogen coupled with fuel cells will even match up economically before 2019, when you consider indirect costs of incumbent technologies and what I dread will be the unavoidable—and possibly severe—price volatility in fossil fuels.

The U.S. saw wild gasoline price fluctuations in 2008. Is it a sign of worse to come? This has been the age of oil, and we might be seeing the beginning of the end of that era. It may be too soon to tell, but world oil production may have peaked at about 87 million barrels a day. If that is truly the top plateau, soon a big shift downward in oil projections will follow. Not many estimating reports appreciate the impact that peak oil will bring. The U.S. Government Energy Information Administration (EIA) long-term predictions are likely too optimistic. Only time will tell.

Realistically, renewable hydrogen technology will not happen overnight, nor will it happen without other changes to the energy source mix. Therefore, we need other energy diversity changes, but should also act to reduce fossil fuel dependence. Under the Obama administration, when you consider that fossil fuels will likely encounter increasing carbon restrictions, whether from cap and trade or from carbon taxes, energy diversity makes sense.

Starting from the reprised base-case 2007 U.S. Primary Energy Consumption Chart on the next page,[8] energy supply source diversity would likely offer strengthened sustainability, enhanced energy security, reduced environmental footprint, improved flexibility to supply disruption, and broadened ability to adapt to new alternatives. The figures show that there is much work to be done to reach a supply source diversity less dependent on fossil fuels (Oil 39.2% + Natural Gas 23.6% + Coal 22.8% = 86.2% energy from fossil fuel sources).

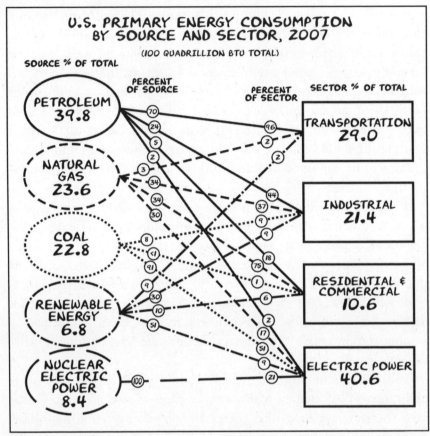

Figure 10.2—EIA 2007 Chart Energy Consumption

(Courtesy of Energy Information Administration—Annual Energy Review 2007, Report No. DOE/EIA-0384(2007), *U.S. Primary Energy Consumption by Source and Sector*, Figure 1)

When you look at energy diversity and the figures above for the first time, you could easily jump to the conclusion that we have too much diversity—that we actually need energy consolidation and simplification. However, when we consider that an 86% market share of fossil fuels is too lopsided and unhealthy for the planet and the economy, and does not fulfill our call as good Christian stewards, we realize we need to expedite change.

We will look extensively at where our energy consumption stands, as taken from the EIA reports. For each of the main energy sources today—namely petroleum, natural gas, coal, renewable energy, and nuclear—we have opportunities to change the mix to something that better fits biblical stewardship and takes better care of God's creation. Remember, the overall demand for energy continually grows, so even if we reduce the percentage of energy by supply (the numbers in the left side oval in Figure 10.2), the total amount used might stay the same or even go up.

For hypothetical example, if we receive 50% of our total energy source from petroleum and we consume 100 billion gallons of gasoline equivalent in year 2009, then our total energy consumption is 50 billion gallons. Now say that in year 2012, we have reduced our petroleum percentage to 45%, and we now use the total energy equivalent of 120 billion gallons across all fuel sources; then a total of 54 billion gallons of gasoline equivalent would come from petroleum. (120 billion gallons x 0.45 = 54 billion gallons.) Therefore, in year 2012 of our hypothetical example, we use 4 billion gallons more than we used in 2009, although our percentage of oil usage dropped 5%. We face a daunting task to reduce total fossil fuel usage, so where do we go next?

It makes sense to tackle the most lopsided energy sectors first. Figure 10.2 shows the United States' energy consumption mix in 2007. It doesn't change drastically from year-to-year. It's up to us to change that. The most glaring deficiency problem is petroleum's 96% monopoly in transportation fuels, which devours 70% of the oil supply. That oil crutch will snap someday.

It's up to us to diversify to avoid collapse—and nowhere is it more important than in the *transportation sector*. Diversification of transportation fuels involves temporary maneuvers and long-term solutions. When scrutinizing potential solutions, especially under the microscope of good stewardship, fuel cells supplied with renewable hydrogen invariably remain the best answer. Electric plug-in vehicles take second place, in my opinion, but because of the existing liquid-fueling infrastructure, the first transitional fuel source is bio-fuels.

Transportation Diversity from Bio-fuels

Ethanol and bio-diesel can fill a vital role to diversify our fuel supply. Their main benefits—liquid, proven, and presently available—outweigh the shortfalls of bio-fuels as detailed in Chapter 8. Bio-fuels provide an avenue away from oil *today* that electric, natural gas, and hydrogen cannot provide in short order. Increased domestic oil exploration cannot fill the void any sooner, either. As long as we don't compromise food supplies, and the industry does not expand too quickly (as we experienced in 2007 and 2008), bio-fuels can play a supporting role. Cellulosic ethanol will take a while to develop and is the only ethanol technology that could reasonably supply a good percent (likely no more than 30%) of our transportation fuel needs.

However, even if cellulosic ethanol production conquers its economic and engineering hurdles, the amount of land area required, the effects on water supplies for growing the energy crops, and its effects on food prices and supplies will keep it from the "final answer" ranks. Likewise, bio-diesel has feedstock limitations (although it has more feedstock versatility than ethanol), performance limitations in cold climates, and storage duration limitations. Like corn-based ethanol, it cannot be mixed with its fossil fuel counterparts and transported via pipeline or tanker, but instead must be mixed at or near its point of use. Large-scale implementation would require extensive infrastructure that would likely never supplant the need for fossil fuels. Nevertheless, bio-fuels deserve a spot on the bench for energy diversity, since they reduce dependence on oil, especially foreign oil imports. The federal government proposed a mandate for 35 billion gallons of ethanol by 2017. Since we can only produce about 16 billion gallons from corn-based ethanol, the rest will have to come from cellulosic ethanol.[9] Considering that today we use over 142 billion gallons of gasoline a year (2007) in the United States, one can see ethanol isn't the silver bullet.[10]

On the plus side, bio-fuels provide local income, are closer to carbon neutral than gasoline and petroleum-based diesel, and fit the present liquid fueling structure easier than other options, providing for immediate impact.

Transportation Diversity from Efficiency and Conservation

The next player in transportation fuel diversity is diversification by negation. Conservation and efficiency are "un-fuels." Improving the fuel economy of internal combustion engines, increasing mileage by reducing vehicle weight by using carbon-fiber bodies, hybrid vehicles, and more efficient traffic control all reduce the amount of fuel consumed.

Avoiding superfluous travel of all types, running errands during reduced traffic times, better planning of driving routes, and combining trips also help conserve fuel. Proper vehicle maintenance, including proper tire air pressure, engine tune-ups, new air filters, and removing unnecessary vehicle cargo, can also improve mileage by well over 10%.[11]

In air transportation, more efficient jet engines and optimization of air traffic routes would reduce jet fuel consumption significantly. Similar to using lighter weight materials in vehicles to improve efficiency, reducing airliner weight (without reducing safety) improves fuel economy considerably.

These improvements result in better efficiency and conservation of fuel, but they can only go so far to break our dependence. Oil is still the supply source, and the clock of dwindling supply is still ticking. It buys time to find more permanent answers, as long as we don't use the savings to just drive more miles. The diversity factor here is in reduced pressure on supply.

Transportation Diversity from Natural Gas

Oil tycoon T. Boone Pickens, one of the most interesting individuals on the national energy scene, commands respect from a large audience who would typically dismiss any shift from the present energy paradigm. Becoming one of the first "good ol' boys" to run at making a profit in the alternative energy sector sets him apart. Unlike many in environmental circles who hold more socialistic views, T. Boone is truly capitalistic. When asked in an interview on *Nightline* if he minded being called a "green profiteer," Pickens replied, "I don't think making a profit is something you should need to apologize for."[12]

Pickens plans to diversify natural gas away from electrical power generation use and use it for the transportation sector instead. This plan definitely has some merit and an encouragingly aggressive timetable, although it may be impractical. Many hope Pickens succeeds. If he does, at least it will familiarize the public to gaseous vehicle fuels.

215

Pickens' plan also recognizes that natural gas works only as a transition fossil fuel to a better gaseous fuel: hydrogen.

According to the 2007 EIA chart, the transportation sector uses only 2% of natural gas, while a third of the natural gas supply goes to provide 75% of the energy in the residential and commercial sector. Natural gas supplies a 30% share of electrical power by source allocation, which equates to 17% of the electrical generation sector. Mr. Pickens wants to displace that 17% with wind power, making natural gas readily available for transportation sector use according to his initial plan.[13] It provides an interesting diversity shift, although I think we can move directly to hydrogen and largely skip the natural gas for transportation step.

Natural gas, while more abundant than oil, still has limited supplies which are estimated to peak in production not far behind oil, according to some in the industry.[14] Remaining conventional supplies are unfortunately found largely in the Former Soviet Union (FSU) and the Middle East—hardly a relief from imported oil problems. However, the diversity of natural gas provides a useful temporary benefit; as the cleanest burning of the fossil fuels, it minimizes carbon emissions. Plus, natural gas provides a transitionally cheap method to generate hydrogen from steam methane reforming.

Another technology to consider for increasing energy diversity is transforming gas to liquids. The most popular is liquefied natural gas (LNG). This allows for the transportation of higher volumes of natural gas in liquid form for long distances. This increases the available supply and the practical usage of natural gas to places it would not otherwise be able to go, from the Middle East and FSU countries to the United States, Japan, and other countries. It requires about 30% more natural gas at the raw supply point to compress it and liquefy it, and it requires more energy to expand it back to gaseous form at the other end of transportation. But this would be much better than wasting natural gas by flaring it in oil fields as a waste product.

Transportation Diversity from Electrical Power

Plug-in, all-electric cars will likely be the closest competition for hydrogen fuel cell vehicles. As discussed in Chapter 6, the two technologies complement each other; neither is truly an energy source, but an energy carrier. Nevertheless, all-electric vehicles do have one advantage at the moment, because the electric grid already exists, whereas the centralized hydrogen fueling infrastructure does not.

That being said, all-electric vehicles have four drawbacks which I believe will prevent them from supplanting hydrogen fuel cells in the end:

- Charge times will never be as fast (at a reasonable household voltage level) as refill times for hydrogen
- Electricity does not store as well as hydrogen, especially as energy output size increases for batteries
- The electric grid is aging and in need of major overhaul to keep up with normal growth, much less the added load from electric car charging
- And lastly, the all-electric vehicle range and battery weight will likely not compare with the technical performance of fuel cell systems when they reach maturity.

Electricity's role in the hydrogen energy economy will be large, though. To kick-start the hydrogen economy, off-peak, base-loaded electricity generation can power electrolyzers to make, compress, and store hydrogen for transportation fuel use. Longer term, electrolysis based hydrogen generation must grow to replace fossil fuel based hydrogen generation. Of course, this means we need to move from fossil fuel electricity generation to renewable electricity generation. Why not use electricity directly? The reason lies with electrical storage issues—answers that hydrogen can provide. The pairing of hydrogen fuel cells and batteries in vehicles makes a good marriage of technologies, too. The energy-carriers electricity and hydrogen are more like squabbling sisters than mortal enemies. The success of one will benefit the other.

Transportation Diversity from Nuclear

It doesn't seem right to mention "nuclear" and "transportation" in the same sentence, but it is legitimate. Large-scale, base-loaded nuclear power plants now exclusively produce electricity. Electrolyzing water in off-peak hours when the electricity would be wasted anyway could be one simple way to diversify and generate hydrogen for transportation or other use.

In the future, nuclear plants should also produce bulk hydrogen via high temperature electrolysis or thermal-chemical water-splitting in addition to electricity. Since electricity cannot be effectively stored in bulk, stored hydrogen created by nuclear power would effectively increase nuclear power's diversity and provide a new avenue into the transportation fuel sector without emitting any greenhouse gases.

The future success of nuclear energy resides with high temperature Generation IV technology commercialization. However, even for presently supported Generation III designs, the regulation and resulting time delays lead to amazing slowdowns in actual plant construction and operation.

In the United States, it's been more than thirty years since the last new nuclear plant went through the full process to commercialization, so the odds of building more than 5 to 15 new reactors by 2030 are slim. And the desire to build 45 to 100 more reactors in the United States, as some have proposed, remains logistically, economically, and socially impractical. The regulatory climate has improved for the moment, but could turn on a dime with any political shift or radiation incident. Without federal government finance guarantees and insurance backing, most utilities cannot afford the risk premium. Soft and uncertain political support, combined with lengthy project development times, make nuclear projects financially unattractive. Something will get built in the United States, and giving nuclear power the benefit of the doubt, it's included as a net scope increase in the 2030 estimate in Figure 10.3 on page 221.

Transportation Diversity from Renewable Energy

Outside of bio-fuels we already discussed, renewable energy technologies supply electricity that empowers production of hydrogen through electrolysis. Growing the supply of renewable power could help produce the hydrogen we need for both central and distributed sources.

But the fact that only 6.8% of the country's energy source comes from renewable energy (mostly hydroelectric) is a great stumbling block. Looking at the data from the energy sector side proves disheartening also. Only 9% of electric power generation, 2% of transportation, 9% of industrial, and 6% of residential and commercial comes from renewable energy.

Coupling these intermittent renewable energy technologies with energy storage would make a big improvement in diversification. Without this type of diversity, the future of renewable energy will be severely limited, possibly to only 20% of electrical generation because of the destabilizing effect it has on the electric grid. However, with the coupling of energy storage to hydrogen or some other form, the power output becomes both more reliable and more available.

Solar is back in vogue again, and although most present technology and manufacturing of solar PV and solar thermal occurs overseas, an opportunity to "skip a generation" to the new CIGS and other technologies could catch us up quickly to the rest of the world in terms of the production and manufacturing of solar cells. Rooftop solar arrays could give a boost to home electrolysis units for hydrogen production.

Off-shore wind power provides an untapped resource when coupled with wave and tidal power, and could produce immense amounts of renewable energy near coastlines, where many live. Wind power in the Midwest enables the United States to become, "The Saudi Arabia of Wind." Renewable energy from geothermal and small hydroelectric provides the United States with ample opportunity to grow electrical output.

Transportation Diversity from Coal

Looking at the source side of EIA Figure 10.2, we see that an astounding 91% of coal supply finds its way to the electric power generation sector, which supplies 51% of the utility electrical power. If coal continues to play a prominent role in electrical power, which it most likely will, a serious expansion in clean coal technology must occur. But the question for the transportation sector is coal's role in transportation fuels. There are a couple options, but unless other methods fall short and need a boost, I don't see how coal will become a major producer of transportation fuel without first being used to generate electricity.

Coal could supply some of the liquid fuels, like gasoline, if absolutely necessary. Coal-to-liquids (CTL) processing, sometimes known as the Fischer-Tropsch process, provided fuel for the German military in World War II and for South Africa during the apartheid years. This last-resort option doesn't solve any of the long-term transportation fuel problems, and reduces coal available for electricity generation and other uses. That's why it has only been produced in large quantities during times of extreme supply shortages. Future projections in the 2008 EIA Annual Energy Outlook show greater expectations in biomass-to-liquids (BTL) planned capacity. But if hydrogen and fuel cells take off better than expected by the EIA, neither CTL nor BTL will be needed.

The other possibility is coal gasification to produce hydrogen, while sequestering carbon dioxide. This has merit as a transportation fuel, but is not the first choice. However, because the powerhouse and developing powerhouse nations such as the Unites States, China, and India have much bigger coal reserves than oil reserves,[15] coal could be pressed into service for transportation fuel rather quickly, especially if the Middle East plummets into Armageddon.

Diversity Roundup

Since 86% of our energy resources come from unsustainable fossil fuels, the United States must actively plan for a world without cheap fossil fuel. As demand grows, world population explodes, and supply tightens, this innocuous-looking EIA chart will represent a dangerously unsustainable world, if left unchecked. The immense size of the incumbent energy system makes it even more crucial that we immediately start diversifying to renewable energy coupled with energy storage, to a little more nuclear power, to expanded efficiency and conservation measures, and to a lot more hydrogen and fuel cells.

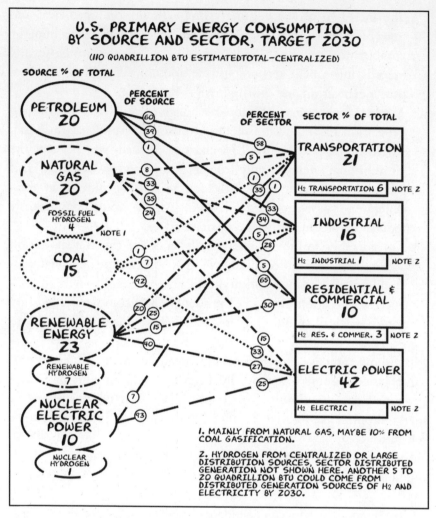

Figure 10.3—Chart of Energy Consumption Targets for 2030 (Format of Annual Energy Review 2007, Figure 1)

Using the EIA chart format from Figure 10.2, an aggressive estimate of where the United States target should stand in the year 2030 is shown on page 221. Compared to the EIA 2008 Energy Forecast for the United States, the total planned energy usage comes to about 7% less than the EIA forecast for 2030 of 118 quadrillion Btu. These reductions could come from a number of sources, including but not limited to the following:

- The increased use of stationary fuel cells in combined heat and power (CHP) configurations helps efficiency, since this technology combination can reach 85% combined efficiency, which greatly exceeds incumbent technology methods of centralized electrical power generation and on-site fossil fuel heating.
- A much higher than EIA forecast usage of renewable distributed generation technologies (see next section) for residential use that would not appear on the EIA chart as defined, which keeps the conventional energy supply from needing to grow as much. This includes running more vehicles on residential-unit, renewably-generated hydrogen than planned, which would significantly reduce petroleum prime energy demand.
- Considerably improved energy efficiency in the transportation sector, because nanotechnology lightweight carbon-fiber materials could drop fuel consumption drastically.
- Success in cost-effective LED residential and commercial lighting could drastically reduce required electricity demand.

The above can have a huge impact by greatly increased efficiency and conservation numbers, and would keep the total amount of energy consumed to a minimum, which puts less stress on unsustainable energy supplies like fossil fuels, allowing them to last longer for future generations.

The benefits of hitting these year 2030 targets in the chart above would be:

- Tremendously reduced dependence on oil (especially foreign oil) for transportation use, plus increased national security and increased transportation flexibility.
- Reduced carbon dioxide emissions from reducing coal-fired power plant electricity from 51% to 33% of sector energy. Additional carbon reductions from oil on the transportation side, from 96% to 58% of sector energy, will also help combat climate change.

Distributed energy

In order for renewable and alternative energy to succeed, the large-scale centralized electrical power plant philosophy must shift to smaller-scale, localized distributed generation (DG). Hand-in-hand, combined heat and power (CHP) makes sense to include in the new alternative energy paradigm.

Likewise, centralized liquid transportation fueling structures must change to local distributed generation of hydrogen, at least to make the transition. Home fueling stations will become popular and the need for a full-blown centralized hydrogen generation plant and pipeline structure alleviated until further down the road. But with the economic collapse of 2008, an opportunity exists to build a brand-new hydrogen infrastructure, similar to the national highway projects championed by Eisenhower, or the Great Depression infrastructure building projects of FDR. In times of great turmoil, we often have times of even greater opportunity.

Since local electrolysis of water will be one of the contenders for generating hydrogen—and ultimately the cleanest, if coupled with renewable generation of electricity—the electric "smart grid" infrastructure must be expanded greatly, also. This also plays well with net metering of distributed generation: selling power back out to the grid and relieving the need for new central power plants.

We're ultimately looking for a model-shift in thinking: to communities and individual homes and businesses supplying their own electrical power, as well as their own hydrogen for fuel cell vehicles. One of the challenges to developing a hydrogen economy is the "chicken and the egg problem" of which comes first, the hydrogen fueling stations or the vehicles? With distributed generation of hydrogen on-site at home, it's estimated that 78% of most day-to-day driving needs within forty miles are satisfied. That reduces the immediate need for more extensive public hydrogen fueling station development. If current fueling stations incorporate hydrogen stations at the same site and generate their own hydrogen, the industry could potentially jumpstart in short order until quantities of fuel cell vehicles increase. Prioritizing hydrogen fueling stations on major highways would alleviate most of the external fueling station concerns.

To make early adopters of the technology feel more comfortable, the vehicle manufacture's or other parties could provide emergency hydrogen auto club service across the nation, minimizing the fear of being stranded. With GPS systems and satellite to vehicle communications, finding the nearest hydrogen station and calling for assistance would be standard features. Between the convenience of home fill-up and the use of the home power station to provide electrical power, Americans would regain a feeling of independence and cash in the wallet. Imagine seeing your electric utility bill and receiving money from them for the electricity you sold out to the grid! Plus having zero dollars charged for gasoline on the monthly credit card bill! This dream could become a welcome reality if we use renewable energy to create hydrogen and electricity. There's another question that comes up often with distributed generation. Why would the big utility companies, oil companies, auto manufacturers, and other stakeholders in the status quo allow distributed generation to develop? It would seem in their best interests to squash any change from the present infrastructure. Fortunately, they are awakening to the reality that conventional wisdom will not suffice.

For instance, oil companies are not dummies. Petroleum producers understand, even if they won't publicly admit, that oil reserves growth and optimistically low oil field decline rates will not save them from peak oil occurring. It will happen; it's just a matter of timing and severity. We all know they won't abandon their off-shore oil platforms tomorrow, but they do have a vested interest in becoming leaders in the post-peak oil world. If they become leaders in hydrogen production and equipment supply, they open up a completely new, sustainable business line to profitability. Becoming local generators of hydrogen at their existing gas stations is the least capital-intensive way to develop the early hydrogen economy infrastructure. Bulk hydrogen generation at central locations and pipeline distribution can develop a later date when economies of scale warrant it. In fact, large hydrogen production sites exist already for other purposes near 70% of U.S. population.[16] Getting started with hydrogen could accelerate faster than we think.

Another example is the automobile manufacturers. Chrysler, Ford, and General Motors are struggling, in large part, due to lack of innovation. They got stuck with too many gas-guzzling cars, trucks, and SUVs in their product line, while Honda and Toyota proactively developed fuel-efficient cars and hybrids. Innovative products and more research and development by United States car makers would have served their companies better. They now realize that they need to innovate to high fuel efficiency, flexible fuel, plug-in hybrids and fuel cell vehicles. Their previous pattern backfired. Spending money on lobbying congress to keep CAFE mileage standards from increasing, and then stifling any legislation that might hurt their competitiveness instead led to an innovation black hole.

The big question for the United States auto industry is now, "Is it too late to innovate?" All three are financially unhealthy, but the industry is "too big to fail," requiring serious plans for government help and company restructuring. The "big three" are in jeopardy of becoming the "big two." Sometimes you have to hit bottom to turn around. This would be a good time to make a move to fuel cell and all-electric vehicles with home fueling and recharging stations, thereby stimulating the distributed generation marketplace.

The last example is utility electric power producers. One would think convincing them to generate less electricity would be a hard sell. But considering the costs, regulatory uncertainty, and risk of new power plant development, it makes sense to pursue energy efficiency and distributed generation measures to delay those large capital expenditures. New methods of rewarding utility companies for assisting customers in efficiency and distributed generation are changing the way they do business. My local utility, KCP&L, has been recognized and applauded nationally for making these types of cutting-edge changes in efficiency programs.

Build Deliberately Toward a Hydrogen and Fuel Cell Economy

For all the doubts about whether a hydrogen economy will develop, we actually already have an indirect hydrogen economy. We all know hydrogen fuels the sun. Therefore, we already live in a hydrogen economy, since solar energy sustains life on earth now, and even the fossil fuels we use today were once living matter, nourished by the sun.

225

The combustion of hydrocarbons is really about combusting hydrogen that happens to be attached to carbon. Combusting pure hydrogen produces water vapor. Burning hydrocarbons generates carbon dioxide that produces lasting effects (an average of 100 years duration) in the atmosphere, helping to trap heat in and warm the climate.[17]*

To date, we have found it too inconvenient to separate the carbon from the hydrogen before we burn it. But the consequences of those actions have started to catch up with us in the form of pollution and climate change. When we speak of a hydrogen economy, we're really speaking of using hydrogen differently. Instead of direct combustion, we hope to separate the hydrogen, capture and store the carbon, and use the hydrogen in fuel cells.

The game changer for fossil fuels is separating hydrogen for use in pure form. The beautiful benefit of hydrogen is that water is available in much greater abundance than fossil fuels, and we can generate hydrogen through electrolysis. Add to that the fact that hydrogen can be stored in bulk much more easily than its sister energy carrier, electricity, and now you have the most perfect energy solution this side of heaven. But wait, there's more! The hydrogen used in the fuel cell produces very pure water as a byproduct, which provides for other beneficial purposes as well.

The stinky "but" of moving to generation of hydrogen is the cost, which by estimate ranges from $3.00 to $10.00 per gallon of gasoline equivalent, depending on many factors, but $5.00 per gallon is a ballpark estimate.[18] However, the good news for renewable generation of hydrogen is that the cost of carbon will only go up, as carbon-based fuels deplete and as carbon-based emissions incur market penalties or tax burdens.

Some say carbon taxes and carbon cap and trade systems will never truly account for the cost of carbon, which might be true. But we will pay one way or the other, through carbon pollution to the atmosphere and the ensuing indirect environmental cleanup costs. These indirect carbon costs come in many forms: oil spill clean-ups, environmental damage from not cleaning up oil spills, melting of frozen tundra, (which releases the intense greenhouse gas, methane from hydrates),

---

* Carbon dioxide lifetimes can vary widely, as the multiple atmospheric reaction effects and location affect it greatly. 100 years is a generally accepted average.

destruction of ocean habitat from maxed-out absorption of carbon dioxide (CO_2), and climate conditions likely intensified by excessive CO_2 in the atmosphere.

If nothing else, insurance rates have definitely raised, which in part, ties to intensity of weather related events. Actuaries calculate the risk, and insurance companies make business decisions on these numbers. For the hardcore "numbers don't lie" crowd, what more proof could you want? We are failing our God-given mission to subdue (govern) the earth if we fail to manage these risks. However, we should not be dismayed. It is not too late to turn it around, although some in the "climate hammer" crowd would beg to differ. I think it is important to remember that with God, all things are possible. We take responsibility, but we also take comfort from knowing it is in His hands.

We have a lot to be thankful for in this world. For instance, the price of sunshine, wind, waves, and tides, remain as free as the day God created them. Coupling these intermittent electrical generating technologies with electrolysis for producing hydrogen solves both the non-dispatchable electricity problem and the lack of electrical storage problem simultaneously.

For those who say the hydrogen economy costs too much, we should ask, "What is the price of peace worth?" As population expands and resources dwindle, wars will result over those resources, especially oil, the first fossil fuel to cross over the world production peak. We must put our hope in the Lord, not in the drill. Oil will put us over the barrel first, but natural gas and access to clean water will cause an increasing number of wars unless we proactively move to better solutions.

Places to Go, Things to Do

So, if Christians are to lead in energy and environment, how do we do it? We need to build the house on solid rock, not shifting sand.*

---

* Jesus speaks in Matthew 7:24–27: "Therefore everyone who hears these words of mine and puts them into practice is like a wise man who built his house on the rock. The rain came down, the streams rose, and the winds blew and beat against that house, yet it did not fall, because it had its foundation on the rock. But everyone who hears these words of mine and does not but them into practice is like a foolish man who built his house on sand. The rain came down, the streams rose, and the winds blew and beat against that house, and it fell with a great crash."

We need to build on this earth using materials God gave us, in a manner that God wants us to, and choosing wisely the location where we should build. In other words, *what* to build, *how* to build and *where* to build are the questions to answer from a biblical green stewardship perspective.

God in the Bible has blessed and cursed many building projects. He cursed the Tower of Babel because of our human arrogance, building a tower to reach the heavens as if we were gods. That demonstrates *what* we build is important, making sure our intent is right with God. How we build to power our activities is important. Do we build industries with disregard to what God created? Or should we build in harmony with God's creation? Put that way, the answer is obvious. We should build as God has given us dominion to do so, but without abusing His creation. Too many of the ways we produce power today are too questionable to be considered good stewardship projects. We can, and should, do better.

How we build is also important. God told Noah how to build the ark, told him what materials to use, and provided the blueprints. Although built for a godly purpose, the ark wasn't built overnight. It took Noah more than 100 years to build the ark. What a monumental task for one man's family! In addition, can you imagine the ribbing he took from his neighbors, all because he lived in accordance with God's word? But Noah followed God's instructions on how to build and fill the ark with animals. The ark-building project definitely didn't fit in with the conventional wisdom of Noah's day, any more than it would in today's world. Likewise, building a new alternative energy infrastructure today does not fit with conventional wisdom, but it's a godly project with purpose that, like Noah's ark, won't be built overnight. The great flood of Noah's time—the first human-caused climate change event of the Bible—happened because of man's sin. Will we listen like Noah and build for godly purposes an alternative energy infrastructure to preserve a better future for humanity and the rest of God's creation?

Where we build is also important to God. Jesus told us to build our house on a solid foundation. He is the rock to build upon, just as we should build where it makes sense. We should not destroy the landscape with our power plant projects. Whether it's a coal-fired power plant or a wind turbine farm, it needs to be sited properly. Without the proper place to build, can we really expect God's blessing? We want to build wisely, not like the fool who builds where he shouldn't. But we must also avoid the extreme of building nothing. Some have taken new building development from milder opposition such as NIMBY (not in my back yard) even further to BANANA (build absolutely nothing, anywhere, near anything). With this in mind, let's consider what should be done to build and re-build the biggest natural disaster in America of recent times, New Orleans.

The Big Not Easy

Hurricanes Katrina and Rita decimated New Orleans and the rest of Gulf Coast region with a one-two punch. First, Katrina decked large parts of New Orleans, breeching the levy system that keeps the city from flooding. Rita followed up with a knockout blow to the oil and natural gas infrastructure located offshore, and near Beaumont, Texas.

A short three years later, hurricane Gustav eerily reminded Gulf Coast residents of the power of natural forces. The after-effects still resonate throughout the area and the country, especially concerning the human toll of the storms. The oil and natural gas infrastructure has largely recovered. Stuff can be repaired and replaced, but the human tragedy of New Orleans, Pass Christian, and other communities in the area still reverberate in our hearts and minds.

The nickname for New Orleans, the Big Easy, came from its laid-back, worry-free lifestyle. Those days are gone for New Orleans. It would behoove the residents who return to remember the lessons of Katrina and rebuild wisely. The costs of rebuilding are astronomical, tallying approximately $133 billion.[19] Fortifying the levy system to guard against future disasters isn't cheap either, at an expected $10 billion.[20] Questions about how much the government should do, or not do, abound. Some think government should be doing more; others think government should avoid creating a permanent welfare state. New Orleans already struggled with a large population on welfare before hurricane Katrina.

Other questions remain. For example, is it wise to build back areas below sea level, where building should never have happened in the first place? In places like the Netherlands in Europe, where hurricanes never hit, levies make sense, but hurricane regions such as New Orleans face much greater risks.

Compounding the risk is the loss of mangrove and cedar trees from the coast, which provided a natural defense against storm surge, the most destructive element of most hurricanes. When man removes natural defenses, such as mangroves and cedar trees, in order to build houses and businesses, he assumes risk of disaster. Houses in the lower ninth ward of New Orleans flirted with disaster from the time they were built, with improper designs and foundations unsuitable to hurricane winds and floodwaters. As Christ warned in Matthew 7, the houses of the lower ninth ward were not built on a foundation of rock.

We need to clearly address the questions of how and where to build this time, instead of indiscriminately building. Either people need to build outside of harm's way, or building design and construction must follow strict hurricane codes. Lives lost, communities torn apart, and people scattered will make New Orleans a different place from here on out. It would be unwise for New Orleans to rebuild and act the same as pre-Katrina New Orleans.

However, Katrina created many opportunities that otherwise would not have existed. New Orleans has the opportunity to rebuild green. Brad Pitt and others have poured tons of time, money, and effort into rebuilding New Orleans to better protect the residents and their homes. Nevertheless, we must carefully consider how to build, remembering Jesus' words about building our house on solid rock, not shifting sand. I know this is a metaphor for our spiritual lives, but it works in our physical world, too.

Follow the Green Brick Road

Living in Kansas, I can only imagine having my home wiped out by hurricane winds and the even more destructive storm surge that accompanies it. Although my Kansas home will never be rocked by hurricanes, unpredictable tornado "twister" sister storms typically pack even higher winds, making them a great danger to many parts of the country.

Just ask the folks in Greensburg, Kansas, where 95% of the homes and businesses were destroyed by a massive EF5-rated tornado on May 4, 2007. Because of the great devastation and the small population (only 1,574 residents), Greensburg decided to start from scratch and rebuild "green." All government buildings will meet LEED certification, including the first Platinum-Certified LEED building in Kansas.[21]

All of this goes hand-in-hand with the "can-do" work ethic found in many Midwest communities, especially in times of crisis. I spoke with the mayor of Greensburg on a recent trip with the U. S. Green Building Council of Kansas City. Mayor Bob Dixson spoke passionately about how many of the practices implemented in the rebuilding tie back to early settlement days of the community in the 1800s: common sense things—like which way to orient buildings for the best sun exposure, not wasting resources, and reusing materials—which we stopped doing when energy became cheap, but now need to relearn.

Not many places get to start over from scratch. If it wasn't for Greensburg's unique circumstances, they wouldn't, either. Greensburg is dealing with troubles that, hopefully, few of us will have to face. But if we do, we need to be aware of the difference between problems and predicaments. To paraphrase philosopher Abraham Kaplan: problems can be solved, predicaments can only be coped with.[22] What Greensburg does to improve energy efficiency and reduce greenhouse gases will not fix global warming; however, Greensburg can act to solve the problems in their community and set an example for the rest of the world.

Communities both large (New Orleans) and small (Greensburg) can assist in changing people's attitudes. We're all in this together and must endure. By acting wisely and not giving in to defeat, we can provide a full range of answers. New Orleans illustrates the revisions that should be made to large centralized energy infrastructure, and Greensburg illustrates distributed generation: small, localized, but getting the job done.

As long as we act in love for others and our heavenly Father with regard to energy and environmental issues, we're not victims in this world, but leaders. As the saying goes, "You can be part of the solution, or you can be part of the problem." You choose.

CHAPTER 11

FINAL THOUGHTS:
IT'S NOT ROCKET SCIENCE, IT'S GODLY STEWARDSHIP

"For I know the plans I have for you," declares the Lord, "plans to prosper you and not to harm you, plans to give you hope and a future."

—Jeremiah 29:11

God's Plans

The setting surrounding Jeremiah 29:11 describes the Israelites in exile in Babylon, which is in Iraq. The words in Jeremiah are a blessing amazingly fitting to us today. The United States has been at war in Iraq twice in less than fifteen years. Unlike the Israelites in Babylon, the United States has won the military battles, but our nation is still in exile in other ways. We're in bondage to Middle East oil, whether we admit it or not. Our troops are over there "in exile" because of vast oil reserves strategically important to our economy, not because we possess close cultural ties with the Iraqi people. If we were that altruistic, we'd be in the Darfur region and other places in Africa too—but we're not. They don't have the black gold.

Perhaps that's why the war in oil-less Afghanistan withered while the war in Iraq fired up. If the extremely wealthy ex-Saudi, Osama Bin Laden had not found sanctuary with the Taliban in the lawless tribal regions of Afghanistan and Pakistan, would we be there? Afghanistan only provided the radical extremist kinship and terrorist training ground; the money and 9-11 terrorists (19 of 21) came largely from oil-rich Saudi Arabia.

Unfortunately, our bond with the Middle East depends primarily on the national security need to fight oil-enabled terrorists and stabilize access to oil resources. Our Middle East bond has little to do with compassion for the oppressed people of the Persian Gulf; however, we can break those chains and have a chance to build bridges instead, truly winning hearts and minds.

Jesus didn't win hearts and minds by overthrowing the Romans and re-establishing a Hebrew nation, although many thought the Messiah would do just that as he rode into Jerusalem on Palm Sunday. Less than a week later, he hung crucified for failing to meet those misguided expectations. These were just worldly expectations, though. In no way did Jesus fail. He fulfilled his divine purpose.

In tandem, people of faith need to rise above the expectations of this world. We cannot do what Jesus did; God's expectation for us is to fight the good fight *in this world*.

It is important to fight terrorists empowered by oil on their turf, but our involvement in the national affairs of a traditionally tumultuous region, at least in part, empowers those terrorists and creates new ones. We have to be sure we are fighting the good fight. It is more patriotic to fight to *remove* the root cause and enablers of the wars we now fight, as well as wars our children might have to fight in the future.

A patriot loves and builds a better country. Being a patriot does not necessarily require military sacrifice. If we really want to make progress in the Middle East, we must remove oil from the equation. Even then, it won't be easy, but at least we will have disarmed the powerful weapon of oil. As Christians, we need to work above the greedy expectations of the world, like fighting to control the kingdom of oil. Jesus taught the road to heaven wouldn't be about convenience. It's not about the power of oil; it's about the power of God in this world and the next. The Bible expects us to use our God-given resources wisely.

In following Scripture, we can take heart that the Lord's promise to the Israelites in the Book of Jeremiah holds just as true for us today, since Christ has saved us. All we need to do is follow Christ. As a result, we should demonstrate our love by doing good works that God has prepared for us to do! We have the easy part; Jesus had the hard part. God plans to prosper us, but not unless we read and follow His Word.

For energy choices and the environment, it boils down to this—do the right thing. We know what it is: we should care for the earth God lets us use, while loving God and our neighbors as ourselves. It's not rocket science, it's godly stewardship. We should seek the cleanest, best value, least confrontational, most compassionate energy source in order to follow God's command to us in Genesis 1:28 and 2:15. If an energy source could be described as humble but powerful, that's what we'd want. It seems we're looking for an energy source equivalent to the nature of Christ Himself.

Hydrogen fits the description perfectly. A simple, unassuming atom (unlike the complex hydrocarbons that make up oil and coal), hydrogen is pure, clean, plentiful, and storable. But it's also powerful and deserves respect. It's not cheap, just like the cost of our sin is not cheap. Christ has already paid the price of our sin, and that act of grace results in many blessings in this world and the next. We should accept the gift of hydrogen as another blessing from God.

Surprisingly, the hydrogen economy doesn't mean financial ruin. The direct cost of the cheapest energy sources, such as coal and petroleum, are actually the worst financially when you include their indirect costs. These costs include pollution, national security expenses related to securing foreign energy sources (war costs), environmental cleanup and degradation, health costs due to poor air and water quality, climate change, and global impacts to the poor and developing regions outside our borders. So-called experts could argue for years on what the actual indirect costs amount to, but remember this: it's always more than the price that you see. If you see gasoline at $2.56 per gallon, the actual cost is more; if you see it at $4.11 per gallon, the actual cost is more.

There are many people who believe the opposite when they see higher prices, especially at the gas pump. They suspect price gouging and market manipulation by short sellers. Some of this goes on—it always has—but volatility ties to uncertainty, and uncertainty leads to speculation on indirect costs. Mistrust of oil companies and governments leads people to demand congressional inquiries and consumer group saber-rattling, without any real consequences. Windfall profit taxes leveled on oil companies in the 1980s yielded poor results. Often these inflated prices more accurately represent the inclusion of these ugly indirect costs. Unfortunately, some do unfairly profit from reaping these indirect costs, at least temporarily, such as the run up to $147 a barrel in July of 2008, driven in part by speculators.

Then equally unrealistic to the true value, the drop in oil prices under $70 a barrel flowed from the economic collapse in the credit markets in September, 2008. The mass media and citizens shifted focus to the completely man-made financial crisis and away from the physically resource-limited, vastly more dangerous, looming energy crisis. Are people aware the deflation of oil prices resulted from the economic recession, not from true market correction or newly found supply? In essence, most of us do not understand the true cost of oil or other fuels anymore than we understand the true cost of our sin which Jesus has paid in full.

Following the Star

The wise men followed the star to find the Christ Child in Matthew 2, a trip of dedication and personal sacrifice that took about two years, contrary to the popular manger scene depiction. Today, people of faith need to "follow the star" to sustainable energy and an environmentally balanced world. To be perfectly clear, Christians should practice green stewardship—not because stars such as Arnold Schwarzenegger, Cameron Diaz, Brad Pitt, Leonardo DiCaprio, Al Gore or others say so—but because God says so in the Bible.

Although we may try to attain the same results as secular environmentalists or sustainable energy proponents, deep down we have a different purpose, a different "why." We do it for the love and glory of God; that is enough. But when we seek these things, an abundance of other blessings may come with it if we first seek God's will (2 Chron. 26:5). Benefits such as a cleaner planet, energy security, and even financial savings and opportunities are possible.

So why aren't we already all over it, pushing alternative energy and environmental stewardship? We should be all over it, like glue on a kindergarten art project. Unfortunately, Christian churches react to cultural society in many different ways, and not always effectively, when it comes to alternative energy and environmental efforts. Some churches withdraw from environmental items which seem too liberal or progressive, or based on new age religious principles. Other churches try too hard to mandate change politically and socially, leading to a perception of the church as domineering, pushy, or holier-than-thou, which produces an effect opposite to that desired. Other churches acquiesce to the culture's worldview, attempting to bend the Bible to balance with modern-day life. The Bible warns us, though, about watering down the Gospel message, so extreme caution must be taken to remain true to Scripture.

The best approach exemplifies how Christ operated. Jesus did not withdraw from society. He spent time amongst the tax collectors, prostitutes, and sinners. He didn't force others to follow Him; He made it clear they had to choose to follow Him. He didn't bend the Word of God to fit the wishes of society, He didn't perform miracles to please the crowd or show off for His disciples, but strictly for the glory of the Father. Christ went out into the world while staying rooted to the Word of God, boldly teaching but not forcing the will and Word of God on the individual or the masses. God created us with free will, and Jesus never once forced His divinity on anyone.

The fine line that links godly stewardship and alternative energy and the environment must find a similar balance. We must not fall into the trap of worshipping the creation over the Creator. We must not withdraw from the world, refusing to talk to secular environmentalists. We must stay connected to show through our actions that Christians care, too. We must stay connected to environmental causes and alternative energy promotion, even if it means rubbing elbows with politicians, big business, tree-huggers, and other sinners. (After all, we are sinners, too.) We must stay connected to the Word of God and seek its truth, and not succumb to the ways of the world. No matter how convinced we are that biblical environmentalism is correct, we must not force it on anyone, but instead help people come to that conclusion themselves with the help of biblical precepts.

Slavery

Some see energy resources (especially oil) as the slavery of modern times. It stands to reason that because of sin, energy and green stewardship will never fully break away from the slavery of fossil fuel. Other

Christians would argue the way we currently use the environment and energy resources is not a sin at all. They claim the world is ours to use as we see fit, since God made us rulers over earthly creation. Therefore, there's nothing wrong with the incumbent energy system; it is acceptable practice. They seriously doubt recent climate change stems from human causes, and largely believe the pending shortage of energy supplies, especially oil, is a hoax. Since God provides, we should have faith and not worry. We therefore should accept the energy infrastructure that God has allowed to develop and dismiss any claims that the energy system is enslaving us, or that we are enslaving the environment. Frankly, they just do not believe it is a priority for furthering God's kingdom.

However, a growing number of Christians disagree with that approach. They claim the present treatment of the environment and energy resources is sinful. Therefore, we should not just accept that "this is just the way it is." In their eyes, we're falsely enslaved to the world, to energy resources from foreign lands, and to the pollution caused by our hand. These Christians also believe the shackles that enslave us to cheap and unsustainable energy sources are a hindrance to God's kingdom.

Through history, the imperfect sinful world we live in clouds our judgment and reminds us of our sinful condition. The slavery we suffer to our present energy infrastructure bears resemblance to human slavery. The institution of slavery existed in the Bible in both the Old and New Testaments. Slavery existed in almost all societies, some of which kept more than half of the population in chains. Slavery seems to be accepted, but not condoned, in many instances in the Bible. As recently as the United States Civil War, biblical scholars argued both for and against slavery, each claiming Holy Scripture gave them the moral high ground.

The New Testament book of Philemon, written by the Apostle Paul, dealt with the return of a runaway slave, Onesimus, to his owner, Philemon. Onesimus, by law of the society at the time, deserved torture and death for running away. Nevertheless, Paul pleaded for grace and mercy for the newly-converted Christian Onesimus, and sent him back to Philemon. Paul did not condemn or condone the institution of slavery in his letter to the wealthy slave owner Philemon in the first century of Christianity, but instead worked within the society at hand to encourage a new doctrine of acceptance as equal brothers in Christ (see also Galatians 3:28: "There is neither Jew nor Greek, slave nor free, male nor female, for you are all one in Christ Jesus").

In our slavery to incumbent energy technologies today, we should act in much the same way, neither condoning nor condemning the entrenched energy infrastructure. Nor should we condemn nor condone other Christians who disagree with us. We should act as Paul encouraged, and work from within the energy infrastructure in order to change it. We should also work with Christians who don't agree with us, secular environmentalists, heads of government, and any person we hope to see some day in God's heavenly kingdom. It is the realistic response and the only course of action that will someday lead us to a society dominated by sustainable, hydrogen-based energy. This frees us up to serve God's kingdom and lead others to heaven by our loving and caring actions.

The Greatest Commandment—The Law of Love

> "Teacher, which is the greatest commandment in the law?" Jesus replied, "Love the Lord your God with all your heart and with all your soul and with all your mind. This is the first and greatest commandment. And the second is like it: 'Love your neighbor as yourself.' All the Law and the Prophets hang on these two commandments."
> —Matthew 22:36–40

What happens when we lose hope? Is there anything more damaging to the human soul? Without hope, the depravity of spirit decimates the body and soul, a terrifying hole too deep to escape without a "whole" God. Fortunately, we have a whole God, in Christ, who can save us from our self-dug hole.

As the saying goes, "When you've dug yourself into a hole with your problems, the first thing you should do is put down your shovel." Unfortunately, we haven't put down the shovel yet and continue to pursue business as usual, digging for coal and drilling for oil. But if we at least slow down our digging and drilling for a minute, we can find ways to better use fossil fuels, as well as improve nuclear technologies until the day when renewable technologies and fuel cells dominate the energy market.

Unlike the hopeless cries you hear in the mainstream media, Christians can relax in the arms of a loving God, blanketed in hope. But we should only release our worries. God still demands action on our part, to live up to our inconvenient purpose of honoring God, respecting His creation, and caring for others.

Without God's love, we are nothing, but with His love, we have hope, grace, mercy, and peace. We must renew our commitment to good stewardship because of our responsibilities and relationships to God and our neighbors. The Bible says that our neighbor is everyone, both believers in Christ and non-believers, as Christ taught us in the Parable of the Good Samaritan (Luke 10:25–37). The call of the greatest commandment leads us to a greater sense of stewardship, for showing strong stewardship embodies action, not words.

Since stewardship embodies action and not words, it provides visual and tangible evidence of caring in action. Good stewardship embodies our Christian responsibility to show the law of love expressed in the Great Commandment. Show your love by the actions you do to bring honor and respect to God in heaven. Care for the creation with which God has abundantly blessed us.

Love the Lord by seeking new ways to renew your commitments to godly stewardship. Our non-Christian neighbors are watching. Demonstrating love and respect for God's creation is a wonderful way to reach out to those neighbors. They understand there's a moral or spiritual side to protecting the environment. It's up to us to convey the love and respect we have for God through our good stewardship of the environment. When we don't convey God's love, it creates an obstacle to them accepting Christianity.

You don't have to be a Christian to care for the environment, so we as Christians should do everything we can to remove the stumbling blocks of bad stewardship. It's not enough for us to care only for our neighbor's spiritual relationship with God; we must also care for their physical world also. If we give a person food, but destroy their land, how many souls will be saved? If we claim to follow Christ, but pollute and waste resources, will non-Christians want to worship Jesus and be like us? To be a good steward emulates the love God shows us, and in turn shows God's love to our neighbor. The cross emulates the Great Commandment, the law of love. The vertical represents our relationship with God; the horizontal, our relationship with our neighbors.

None of this care for the environment, the entanglements of our energy situation, contradicts the teachings of Jesus. While His concern rests primarily with the spiritual (our relationship with God), that does not relieve us from the obligation to be good stewards of His creation. We can only progress so far with conventional technologies such as combustion of fossil fuels, conservation, and energy efficiency. The road out of poverty leads to a more energy intensive lifestyle, which further stresses conventional technology fuelled by fossil fuels.

World population has expanded tremendously and continues to grow, making conservation helpful, but ineffective in reducing total fuel usage. Physical and practical limits restrict efficiency improvements, leading to ineffectual control of fuel consumption. Even if conservation and efficiency work temporarily to reduce fuel usage, the price temporarily drops, which usually has the rebound effect of consumption increasing again. That's why we need to break from the conventional wisdom of the day and trust in God's provision, not by hardship, resource wars, and restricted prosperity, but by innovative, sustainable, environmentally safe new technologies available to all.

Hydrogen and fuel cells fulfill all these requirements. The so-called high cost of the hydrogen economy is not an issue if you consider the indirect costs of our present energy structure. Present-day price volatility of energy cascades quickly to high food prices, making them particularly economically disruptive to the poor. Dangerous inherent costs also lurk in the un-captured economic costs to national security and the environment, and the spiritual slippery slope of relying on so much oil from Muslim lands. With hydrogen and fuel cells, we can have local, renewable, clean, sustainable, safe, and independent energy.

Economics, energy, and environment depend on each other. We cannot address economic and energy issues without addressing environmental impact, and preserving our environment should not force us back to the Dark Ages, before the days of the industrial revolution, electricity, and high-speed transportation. We find balance by first trusting in God. He made the world much more resilient than we understand. God loves not only us, but all of His creation, and ultimately, He has the final word.

Next, we need to build up distributed energy supplies and technologies. This will relax the requirement to build more power plants and refineries. It will make us more self sufficient, promote national security, and remove spiritual stumbling blocks to our faith. Next up, we need to embrace alternative energy sources, increasing energy diversity. We should embrace technologies which solve more problems than they create.

Hydrogen fuel cell technology holds the most promise in this regard. Hydrogen stores energy in bulk much better than electricity, and it does not pollute as burning hydrocarbons for power and transportation do. Hydrogen is available from multiple sources, and fuel cells are versatile and scalable. They mate up well with other renewable technologies like wind and solar, which are intermittent and require energy storage/retrieval systems to stabilize power output. They remove the variable of unsustainable resources from the environmental and energy equations. They break the bonds of dependence on non-Christian, terrorist-harboring governments. They free us to be good stewards of everything Jesus has taught.

Good stewardship doesn't mean waiting for all the answers and then taking action. Wise decisions can be made without all the facts, without the need to wait for scientific proof that may never come. Even if it did, the timing would be way too late to make a difference.

On the other hand, we should not fear and fret about the future, either, if we are not living in the moment, enjoying the God given gifts we have. Jesus warned Martha of the importance of the here and now when Mary spent her time with Jesus rather than helping the worried Martha with housekeeping chores. We cannot focus on the needs of others if we're too preoccupied with drilling for oil and tunneling for coal. Instead, we need to focus on helping the poor at hand and dreaming up new paths to help them, which in turn helps ourselves.

It is not fiscally responsible to expect the status quo to end the debate through science over the next 100 years; wiser, more prudent ways demand implementation here and now. And like the parable of the talents, we need to put the new ways to work now, not bury them for another ten years waiting for easier political and economic times, or peace in the Middle East.

As taught in the parable, we know God is a demanding master. He expects us to grow His kingdom in every way possible. Bowing to the oil purveyors of the Middle East, ignoring the environment, and propagating unsustainable energy infrastructure all fail to fulfill the demand God has made. God expects us to accomplish the difficult, not avoid it.

The financial and credit meltdowns unveiled in late 2008 present us with great challenges, but also great opportunities. God has blessed us with the opportunity to build a new energy infrastructure and industry, placing the economy and society on the right course. Christians have an opportunity to design, lead, and build a sustainable road aligned with biblical green stewardship.

God called us to "fill the earth and subdue it" in Genesis 1:28. We've done a great job of filling the earth with 6.5 billion people, but we have not done as well "subduing," or governing the earth. Nothing in the Bible has released us from that decree.

Christian Living

How should Christians respond to non-Christians on energy and environment issues? First and foremost, Christians need to demonstrate that they indeed care for the environment and also desire to be good stewards of the resources God has given us because the Bible, the Word of God, says so. This means action, not just words. Secondly, Christians need to be clear on the reasons why we love the environment and resources that God's creation provides. Namely, we do it out of love and respect for God, and love for our fellow man. We should always have a healthy, fearful reverence of God's power and might. Thirdly, we must emphasize that we do not do this to earn God's love or bring God's favor upon us by our own actions, or to buy our way into heaven. No, we must be clear that we do these things because of

God's love, grace, and mercy on us and not in an attempt to earn or buy God's grace. We reflect God's love by loving what He loves.

How should Christians respond to fellow Christians panicked over energy and environmental issues, or those on the other end of the spectrum, who believe energy and environmental realities are a hoax?

For the first group, the worriers, remember Matthew 6:25–34, where Jesus tells us not to worry, but to seek God first, and the rest will fall into place. Reinforce that it's all in God's hands and He is infinitely more powerful than any potential tragedy. Take comfort in the fact that no matter what we do, this world is temporal, so as long as we practice the law of love and renew our call to be good stewards of God's creation here on earth, everything will work out.

The supremacy of Christ guarantees everything will work out. Colossians 1:15–17 states: "He is the image of the invisible God, the firstborn over all creation. For by him all things were created: things in heaven and on earth, visible and invisible, whether thrones or powers or rulers or authorities; all things were created by him and for him. He is before all things, and in him all things hold together."

For the second group—the ones who believe that God provides, the ones who say we need not worry about the hysterics of global warming fanatics, the ones who tune out whenever they hear Al Gore's name uttered—remember that God's Word is truly amazing! The same verses from Matthew 6 provide the answer. This time, note that the exhortation to not worry does not mean do nothing, but rather act today in a righteous manner and the future will take care of itself.

Too many Christians misinterpret "do not worry" as an anthem for the status quo. They believe we can help the poor and live our lives as we've become accustomed to, keeping the same business models and problematic energy infrastructures in place. Albert Einstein and others have wisely noted that what gets you into a problem rarely solves the problem.

I guarantee that more oil and more coal is not the answer. Better use of oil and coal will be part of the solution, but definitely not the final answer. Christians need to renew their calling to good stewardship today combining conservation, energy efficiency, and most importantly, leading the charge to sustainable and cleaner energy choices, such as hydrogen fuel cells. We might not be able to change energy infrastructure and the government overnight, but we can change our personal *attitudes* overnight, or for those stubborn folk, at least within a month.

Our attitudes can be about people or about situations. People often naturally take an attitude that those who differ with them are wrong, and seek out those who as agree with them. Sometimes this is the right thing to do, sometimes it isn't. The attitude or barrier that we need to break down regarding energy and the environment, at least politically, include the stereotypes of the wacky left and the heartless right. These divisive stereotypes do nothing beneficial to build God's kingdom. They serve no purpose.

Our attitudes, especially in regards to bad situations, say a tremendous amount about our character. We have only two ways to handle bad situations. Fix it if you can; accept it if you cannot. And for any situation, good or bad, keep your personal relationship with Jesus Christ strong. Always remember to pray.

> Jesus, You are truly the Son of God, the Prince of Peace. We pray that we may deeply realize in our hearts and minds that You have won every battle and war against sin; not with legions of angels, but with Your lone suffering and death on the cross. Risen Lord, Your victory over death has made us whole again. Grant that we may praise and worship You by caring for creation. Let us be a light on a hill, a beacon of light and love and care for all people and for all creation. Lord, your first command to us was to fill the earth and govern it. Let us gently subdue the earth, as Christ has reconciled all creation to the Father's loving arms. Let us be good stewards to everything you have left us to care for with mercy and grace.
>
> And if we fail, grant us the strength to pick ourselves up and try again, always discerning, always seeking to do your will, Lord Jesus. Let us always show love.
>
> In Christ's holy name, we pray. Amen.

ENDNOTES

Preface

1. The American Chesterton Society, "Quotation of G. K. Chesterton," *The American Chesterton Society*, http://chesterton.org/acs/quotes. htm.

Chapter 1

1. Wisdom Quotes: Quotations to Inspire and Challenge, "Change and Growth Quotes: Anne Frank," Jone Johnson Lewis, http:// www.wisdomquotes.com/cat_changegrowth.html.
2. Energy Information Administration, "Energy in Brief" *How Dependent Are We On Foreign Oil? Energy Information Administration:* http://tonto.eia.doe.gov/energy_in_brief/foreign_oil_dependence. cfm (August 28, 2008).
3. Energy Information Administration, "Executive Summary-Figure 2" *Electric Power Monthly February 2009. Energy Information Administration DOE/EIA-0226(2009/02), www.eia.doe.gov/cneaf/ electricity/epm/epm_sum.html* (February 13, 2009).
4. See note 2 above.
5. See note 2 above.
6. See note 2 above.

Chapter 2

1. United States Census Bureau, "Population Finder," *American Fact Finder*. http://factfinder.census.gov/servlet/SAFFPopulation?_submenuId=population_0&_sse=on.

2. Energy Information Administration, "Energy in Brief" *How Dependent Are We On Foreign Oil? Energy Information Administration:* http://tonto.eia.doe.gov/energy_in_brief/foreign_oil_dependence.cfm (October 26, 2005).

3. Energy Information Administration, "World Proved Reserves of Oil and Natural Gas, Most Recent Estimates," *Energy Information Administration*: http://www.eia.doe.gov/emeu/international/reserves.xls.

4. Federal Highway Administration, "State Motor Vehicle Registrations-2006," *Highway Statistics 2006*. http://www.fhwa.dot.gov/policy/ohim/hs06/pdf/mv1.pdf.

5. Energy Information Administration, "EIA Energy Basics 101," *Energy Information Administration*. www.eia.doe.gov/basics/energybasics101.html.

6. Meet the Press. First broadcast 30 April 2006 by NBC. Hosted by Tim Russert. http://www.msnbc.msn.com/id/12518683/.

7. Paul Krugman, 2008, "The Oil Nonbubble," *The New York Times,* May 12, 2008, http://www.nytimes.com/2008/05/12/opinion/12krugman.html?_r=3&em&ex=1210737600&en=4c285a3b0ff54893&ei=5087.

8. David Edwards and Muriel Kane, 2009, "CBS: Did Wall Street Speculators Create Oil Price Bubble?," Raw Story, January 12, 2009, http://rawstory.com/news/2008/CBS_Wall_Street_caused_oil_price_0112.html.

9. National Highway Traffic Safety Administration, "Automotive Fuel Economy Problem: Annual Update Calendar year 2003," *U.S. Department of Transportation*. http://www.nhtsa.dot.gov/cars/rules/cafe/FuelEconUpdates/2003/index.htm.

10. Energy Independence and Security Act of 2007, Public Law 110-140, 110th Congress. (December 19, 2007). http://thomas.loc.gov/cgi-bin/query/F?c110:8:./temp/~c1107Zp6DV:e900:.

11. Center for Media and Democracy, "Lawrence J. Korb" *Source Watch*. http://www.sourcewatch.org/index.php?title=Lawrence_J._Korb.

12. David Wallechinsky, 2008, *The World's Worst Dictators: Top 10 of 2008*, Parade, February 04, 2008. http://www.parade.com/dictators/2008/.

13. Carol A. Dahl, *International Energy Markets: Understanding Pricing, Policies, and Profit* (Tulsa, OK: PennWell Corporation, 2004), 34.

14. Energy Information Administration, "Top Oil Producing Counties & Exporters" *Petroleum Basic Statistics:* http://www.eia.doe.gov/basics/quickoil.html.

15. Energy Information Administration, "Oil Reserves," *County Analysis Briefs: Saudi Arabia:* http://www.eia.doe.gov/emeu/cabs/Saudi_Arabia/pdf.pdf.

16. Matthew R. Simmons, *Twilight in the Desert: The Coming Saudi Oil Shock and the World Economy* (Hoboken, NJ: Wiley, John & Sons, Incorporated, 2005), 27.

17. Ibid.

18. Anthony Cave Brown, *Oil, God and Gold: The Story of Aramco and the Saudi Kings* (Boston, MA: Houghton Mifflin Company, 1999), 83–87.

19. Anthony Cave Brown, *Oil, God and Gold: The Story of Aramco and the Saudi Kings*, 102–105.

20. Energy Information Administration, "World Proved Reserves of Oil and Natural Gas, Most Recent Estimates," *Energy Information Administration*: http://www.eia.doe.gov/emeu/international/reserves.xls.

21. Matthew R. Simmons, *Twilight in the Desert: The Coming Saudi Oil Shock and the World Economy*, 70.

22. Neal Adams, *Terrorism & Oil*, (Pennwell Books, 2002), 72.

23. Energy Information Administration, "World Proved Reserves of Oil and Natural Gas, Most Recent Estimates," *Energy Information Administration:* http://www.eia.doe.gov/emeu/international/reserves.xls.

Chapter 3

1. Julie A. Belz, "Information About Islam, Muslims and the Taliban," Penn State. http://www.sa.psu.edu/muslim.htm.

2. Energy Information Administration, "World Proved Reserves of Oil and Natural Gas, Most Recent Estimates," *Energy Information Administration:* http://www.eia.doe.gov/emeu/international/reserves.xls.

3. Energy Information Administration, "Recent World Supply & Consumption Data," *Energy Information Administration:* http://www.eia.doe.gov/emeu/international/reserves.xls.

4. Central Intelligence Agency, "The World Fact Book: Lebanon," *CIA World Fact Book.* https://www.cia.gov/library/publications/the-world-factbook/geos/le.html.

5. Megan Goldin, "Holy Land's Christians Caught in Middle of Conflict," *The Boston Globe.* April 11, 2006, http://www.boston.com/news/world/middleeast/articles/2006/04/12/holy_lands_christians_caught_in_midst_of_conflict/?page=1.

6. Ibid.

7. Jeremy Rifkin, *The Hydrogen Economy: The Creation of the Worldwide Energy Web and the Redistribution of Power on Earth, (New York: Penguin Group, 2003)* 92, 93.

8. Ibid, 92.

9. Brannon Howse, *One Nation Under Man?: The Worldview War Between Christians And the Secular Left*, (Nashville: B&H Publishing Group, 2005), 13.

10. Jeremy Rifkin, *The Hydrogen Economy: The Creation of the Worldwide Energy Web and the Redistribution of Power on Earth*, 91.

11. Central Intelligence Agency, "The World Fact Book: Saudi Arabia," *CIA World Fact Book.* https://www.cia.gov/library/publications/the-world-factbook/geos/sa.html.

12. Sultan Munadi, "Afghan Case Against Christian Convert Falters," The New York Times, http://www.nytimes.com/2006/03/26/international/asia/26cnd-afghan.html?ex=1301029200&en=c9ed4e6797ef87a8&ei=5088&partner=rssnyt&emc=rss.

13. Meet the Press. First broadcast 16 April 2006 by NBC. Hosted by Tim Russert. http://www.msnbc.msn.com/id/12283802/page/6/.

14. Central Intelligence Agency, "The World Fact Book: Field Listings-Religions," CIA. https://www.cia.gov/library/publications/the-world-factbook/geos/us.html.

15. Central Intelligence Agency, "The World Fact Book: Saudi Arabia," *CIA World Fact Book*. https://www.cia.gov/library/publications/the-world-factbook/geos/sa.html.
16. Central Intelligence Agency, "The World Fact Book: Field Listings-Religions," *CIA World Fact Book*. https://www.cia.gov/library/publications/the-world-factbook/fields/2122.html.
17. Central Intelligence Agency, "The World Fact Book: Saudi Arabia," *CIA World Fact Book*. https://www.cia.gov/library/publications/the-world-factbook/geos/sa.html.
18. Central Intelligence Agency, "The World Fact Book: Saudi Arabia," *CIA World Fact Book*. https://www.cia.gov/library/publications/the-world-factbook/geos/sa.html.
19. David Goldmann, *Islam and the Bible: Why Two Faiths Collide*, (Chicago: Moody Publishers, 2004), 165.
20. Beth Moore, *Living beyond Yourself: Exploring the Fruit of the Spirit*, (Nashville: Lifeway Christian Resources, 2004), 123, 124.

Chapter 4

1. Bill Parks, U.S. Department of Energy, "Transforming the Grid to Revolutionize Electric Power in North America." Paper presented at the Edison Electric Institute Fall 2003 Transmission, Distribution and Metering Conference, October 13, 2003.
2. Energy Efficiency Network News, "World's Largest Utility Battery System Installed in Alaska," *U.S. Department of Energy*, http://apps1.eere.energy.gov/news/news_detail.cfm/news_id=6490.
3. Ann Chambers, *Renewable Energy in Non-technical Language* (Tulsa, OK: PennWell Corporation, 2004), 62.
4. Energy Information Administration, "Electric Power Industry 2007: Year In Review" *Energy Information Administration:* http://www.eia.doe.gov/cneaf/electricity/epa/epa_sum.html.
5. Energy Information Administration, "Existing Capacity by Energy Source, 2007," *Energy Information Administration:* http://www.eia.doe.gov/cneaf/electricity/epa/epaxlfile2_2.pdf.
6. Energy Information Administration, "International Energy Outlook 2008," *Energy Information Administration:* http://www.eia.doe.gov/oiaf/ieo/coal.html.

7. Nuclear Energy Institute, "Resources and Stats: World Statistics," *Nuclear Energy Institute:* http://www.nei.org/resourcesandstats/nuclear_statistics/worldstatistics/.

8. Ann Chambers, *Renewable Energy in Non-technical Language,* 162

9. GENI, "National Energy Grid Canada," *Global Energy Network Institute,* http://www.geni.org/globalenergy/library/national_energy_grid/canada/index.shtml.

10. Mark Clayton, "A Big Wave of Mini-Hydro Projects," *The Christian Science Monitor,* December 19, 2005, http://www.csmonitor.com/2005/1219/p03s02-sten.html.

11. Ibid.

12. Energy Policy Act of 2005, Public Law 109-58, 109th Congress. (August 8, 2005). www.doi.gov/iepa/EnergyPolicyActof2005.pdf.

13 Ibid.

14. Focus on Energy, "Renewable Energy: Wind Turbines and Birds," *Wisconsin Focus on Energy,* http://www.focusonenergy.com/data/common/dmsFiles/W_RI_MKFS_Windturbinesandbirdsv0207.pdf.

Chapter 5

1. U.S Agency for International Development, "Global Population Profile: 2002," International Population Reports, *U.S. Census Bureau,* http://www.census.gov/ipc/prod/wp02/wp-02.pdf.

2. George Ochoa, Jennifer Hoffman, and Tine Tin. *Climate: The Force That Shapes Our World and the Future of Life on Earth* (London: Rodale International Ltd, 2005), 92.

3. U.S Environmental Protection Agency, "Climate Change: Health and Environmental Effects," *Environmental Protection Agency,* http://www.epa.gov/climatechange/effects/polarregions.html#ref.

4. Al Gore, *An Inconvenient Truth: The Planetary Emergency of Global Warning and What We Can Do About It* (Emmaus, PA: Rodale, 2006), 149.

5. Goddard Institute for Space Studies, "GISS Surface Temperature Analysis," *National Aeronautics and Space Administration:* http://data.giss.nasa.gov/gistemp/2008/.

6. National Oceanic and Atmospheric Administration, "Climate of 2008—in Historical Perspective Annual Report," *National Climatic Data Center,* http://www.ncdc.noaa.gov/oa/climate/research/2008/ann/ann08.html.

7. National Oceanic and Atmospheric Administration, "Trends in Atmospheric Carbon Dioxide—Mauna Loa," *Earth System Research Laboratory Global Monitoring Division,* http://www.esrl.noaa.gov/gmd/ccgg/trends/.

8. NASA Goddard Space Flight Center, "The Greenhouse Effect," *NASA Facts Online,* http://www.gsfc.nasa.gov/gsfc/service/gallery/fact_sheets/earthsci/green.html.

9. NASA Earth Observatory, *Clouds & Radiation,* EOS Project Science Office, http://earthobservatory.nasa.gov/Library/Clouds/printall.php, 2.

10. Michael Ritter, "The Physical Environment: An Introduction to Physical Geography," *The Atmosphere,* 2006, http://www.uwsp.edu/geo/faculty/ritter/geog101/textbook/atmosphere/atmospheric_composition.html#Principal%20Gases.

11. Carbon Dioxide Information Analysis Center, "Recent Greenhouse Gas Concentrations," *Carbon Dioxide Information Analysis Center,* http://cdiac.esd.ornl.gov/pns/current_ghg.html.

12. Energy Information Administration, "Alternatives to Traditional Transportation Fuels 1994," *Energy Information Administration,* http://www.eia.doe.gov/cneaf/alternate/page/environment/appd_d.html.

13. Intergovernmental Panel on Climate Change, "Climate Change 2001: Working Group 1," *United Nations Environment Programme*: http://www.grida.no/publications/other/ipcc%5Ftar/?src=/climate/ipcc_tar/wg1/248.htm.

14. Energy Information Administration, "International Energy Annual 2006," *Energy Information Administration:* http://www.eia.doe.gov/pub/international/iealf/tableh1co2.xls.

15. U.S.Environmental Protection Agency, "Emission Facts: Greenhouse Gas Emissions from a Typical Passenger Vehicle," *Environmental Protection Agency:*http://www.epa.gov/oms/climate/420f05004.pdf.

16. Energy Information Administration, *U.S. Carbon Dioxide Emissions from Energy Sources 2005*. Energy Information Administration. www.eia.doe.gov.

17. Sanjeev Ghotge and Ashwin Gambhir, "Global Climate Change: Threat to Nature and Human Society," *Countercurrents*, October 27, 2007. http://www.countercurrents.org/gambhir051007.htm.

18. NASA Earth Observatory, *Clouds & Radiation,* EOS Project Science Office, http://earthobservatory.nasa.gov/Library/Clouds/printall.php, 1.

19. Insurance Journal, "MIT Hurricane Study: Global Warming 'Pumping Up' Destructive Power," Southeast News, August 2005, http://www.insurancejournal.com/news/southeast/2005/08/01/57888.htm.

20. Neil Greenfieldboyce, "Study: 634 Million People at Risk from Rising Seas" *National Public Radio Morning Edition,* http://www.npr.org/templates/story/story.php?storyId=9162438.

21. George Ochoa, Jennifer Hoffman, and Tine Tin. *Climate: The Force That Shapes Our World and the Futures of Life on Earth* (London: Rodale International Ltd, 2005), 94.

22. Vic Camp, Ph.D, *Climate Effects of Volcanic Eruptions*, Project ALERT, www.geology.sdsu.edu/how_volcanoes_work/climate_effects.html.

23. Ibid.

24. Goddard Institute for Space Studies, "GISS Surface Temperature Analysis," *National Aeronautics and Space Administration:* http://data.giss.nasa.gov/gistemp/2008/.

25. Ron Nielsen, *Solar Radiation,* 2005, http://home.iprimus.com.au/nielsens/solrad.html.

26. *Nova. Dimming the Sun,* First broadcast 18 April 2006 by PBS. Directed by Duncan Copp. Produced by David Sington. http://www.pbs.org/wgbh/nova/transcripts/3310_sun.html.

27. Per Strandberg, "The Reason Global Warming is Manmade," Global Warming and the Climate, http://www.global-warming-and-the-climate.com/greenhouse-warming-argument.htm.

28. Energy Information Administration, *Report #:DOE/EIA-0383(2008),* Energy Information Administration, http://www.eia.doe.gov/oiaf/aeo/excel/figure97_data.xls.

Chapter 6

1. Wisdom Quotes: Quotations to Inspire and Challenge, "Change and Growth Quotes: G.K. Chesterton," Jone Johnson Lewis, http://www.wisdomquotes.com/cat_changegrowth.html.

2. Rex A. Ewing, *Hydrogen: Hot Stuff, Cool Science* (Masonville, CO: PixyJack Press, 2004).

3. Rebecca L. Busby, *Hydrogen and Fuel Cells: A Comprehensive Guide* (Tulsa, OK: PennWell, 2005), 403.

4. Office of Fossil Energy, "Fuel Cell Handbook," Department of Energy, http://www.fuelcells.org/info/library/fchandbook.pdf, 1–3.

5. The National Hydrogen Association and The US Department of Energy, *Hydrogen Safety Fact Sheet*, The National Hydrogen Association, http://www.hydrogenassociation.org/general/factSheet_safety.pdf.

6. Kenneth S. Deffeyes, *Beyond Oil: The View from Hubbert's Peak* (New York: Hill and Wang, 2005), 156.

7. Rebecca L. Busby, *Hydrogen and Fuel Cells: A Comprehensive Guide*, 181.

8. Fossil Energy Department of Communications, "FutureGen Clean Coal Projects," Department of Energy, http://fossil.energy.gov/programs/powersystems/futuregen/.

9. Rebecca L. Busby, *Hydrogen and Fuel Cells: A Comprehensive Guide*, 200.

10. Jeff Wise, "The Truth about Hydrogen," *Popular Mechanics*, November 2006, http://www.popularmechanics.com/technology/industry/4199381.html.

11. Rebecca L. Busby, *Hydrogen and Fuel Cells: A Comprehensive Guide*, 38.

12. Kenneth S. Deffeyes, *Beyond Oil: The View from Hubbert's Peak*, 102.

13. Ibid, 56.

14. The National Hydrogen Association and The US Department of Energy, *Hydrogen Safety Fact Sheet*, The National Hydrogen Association, http://www.hydrogenassociation.org/general/factSheet_safety.pdf.

15. Lonely Planet Publications, "Iceland," *International Student Travel Confederation*, http://www.istc.org/sisp/index.htm?fx=destination&loc_id=131056&section=environment.

16. Central Intelligence Agency, "The World Fact Book: Iceland," *CIA World Fact Book*. https://www.cia.gov/library/publications/the-world-factbook/geos/ic.html.

17. Maria Maack and Thomas Schucan, "Ecological City Transport System," *Icelandic New Energy,* http://iea-hia-annex18.sharepoint-site.net/Public/Selected%20Case%20Studies/Ecological%20City%20Transport%20System-ECTOS%20(Iceland).doc.

18. Central Intelligence Agency, "The World Fact Book: Iceland," *CIA World Fact Book*. https://www.cia.gov/library/publications/the-world-factbook/geos/ic.html.

Chapter 7

1. Dr. Lynn White Jr., *The Historical Roots of Our Ecologic Crisis (New York: Harper and Row, 1974), http://www.asa3.org/ASA/PSCF/1969/JASA6-69White.html.*

2. Ken Blanchard and Phil Hodges, *Lead Like Jesus: Lessons from the Greatest Leadership Role Model of All Time,* (Nashville: Thomas Nelson, 2007), 198.

3. FRONTLINE'S Hot Politics. 2007, broadcast by PBS. Directed by Tim Mangini and written by Peter Bull. http://www.pbs.org/wgbh/pages/frontline/hotpolitics/etc/script.html.

4. Marc Gunther, *Faith and Fortune: The Quiet Revolution to Reform American Business* (New York: Crown Business, 2004), 250.

5. Tri Robinson and Jason Chatraw, *Saving God's Green Earth: Rediscovering the Church's Responsibility to Environmental Stewardship* (Norcross, GA: Ampelon Publishing, 2006), 65.

6. John Ortberg, Stephen Sorenson and Amanda Sorenson, *If You Want to Walk on Water, You've Got to Get Out of the Boat* (Grand Rapids, MI: Zondervan, 2003), 69.

Chapter 8

1. Brainy Quote, "Arthur Schopenhauer," BrainyMedia.com, http://www.brainyquote.com/quotes/authors/a/arthur_schopenhauer.html.

2. Alexander Green, "Finding Meaning in the Second Half of Life," *Spiritual Wealth: The Road Map to a Rich Life*, March 2009, http://www.spiritualwealth.com/Archives/2008/20080528.html.

3. Energy Information Administration, "Potential Oil Production from the Coastal Plain of the Arctic National Wildlife Refuge: Updated Assessment," *Energy Infromation Administration*, http://www.eia.doe.gov/pub/oil_gas/petroleum/analysis_publications/arctic_national_wildlife_refuge/html/analysisdiscussion.html.

4. Energy Information Administration, EIA Energy Basics 101, *U.S. Primary Energy Consumption by Source and Sector*, Energy Information Administration, www.eia.doe.gov/basics/energybasics101.html.

5. Ibid.

6. Government of Alberta, "Crude Oil & Sands," *Alberta Canada*, http://albertacanada.com/industries/962.html.

7. Boston World Oil Conference, 2006, "Time for Action: A Midnight Ride for Peak Oil," Boston, MA, DVD 3.

8. Hugh Li and Glen Sweetnam, "Issues in Forecasting Natural Gas Prices," Energy Central Energy Pulse, http://www.energypulse.net/centers/article/article_display.cfm?a_id=762.

9. Energy Information Administration, EIA Petroleum Navigator, "*Weekly US Regular All Formulations Retail Gasoline Prices,*" Energy Information Administration, http://tonto.eia.doe.gov/dnav/pet/hist/mg_rt_usw.htm.

10. Energy Information Administration, Official Energy Statistics, "Glossary," Energy Informaion Administration, http://www.eia.doe.gov/glossary/glossary_r.htm.

11. Energy Information Administration, EIA Energy Basics 101, *U.S. Primary Energy Consumption by Source and Sector*, Energy Information Administration, www.eia.doe.gov/basics/energybasics101.html.

12. "The Myths and Truths of Nuclear Energy," Dan R. Keuter, (March 30, 2006) http://local.ans.org/ne-ny/Keuter.ppt.

13. Ken Park, *The World Almanac and Book of Facts 2006,* (New York: World Almanac, 2005), 760, 846.

14. GM Volt Research Forum, "How did GM Determine that 78% of Commuters Drive Less Than 40 Miles per Day?," gm-volt.com, http://gm-volt.com/2007/12/06/how-did-gm-determine-that-78-of-commuters-drive-less-than-40-miles-per-day/.

15. Energy Star, "Compact Fluorescent Light Bulbs," United States Environmental Protection Agency and United States Department of Energy, http://www.energystar.gov/index.cfm?c=cfls.pr_cfls.

16. Lou Ann Hammond, *Hydrogen—The Longer Tailpipe*, Carlist.com, www.carlist.com/autonews/2004/toyota_fchv.html.

17. Amory B. Lovins and others, *Winning the Oil Endgame*. (Snowmass, Colorado, Rocky Mountain Institute, 2004), 2.

18. Elizabeth Anne Viau, *Find out how the Sun's Energy is used on Earth*, Following the Energy Trail, www.world-builders.org/lessons/less/biomes/SunEnergy.html.

19. Ibid.

20. David Neff, "Second Coming Ecology: We Care for the Environment Precisely Because God Will Create a New Earth," *Christianity Today*, July 18, 2008. http://www.ctlibrary.com/ct/2008/july/23.35.html.

Chapter 9

1. RTE Commercial Enterprises Limited, "World Car Sales to Slow in West," *RTE News*, July 4, 2006, http://www.rte.ie/business/2006/0704/cars.html.

2. Ken Park, *The World Almanac and Book of Facts 2006*, (New York: World Almanac, 2005), 846.

3. Ibid, 766.

4. Mike Milliken, "China Has Nearly 160 Million Motor Vehicles; 76.09% Are Private Cars," *Green Car Congress*, January 3, 2008, http://www.greencarcongress.com/2008/01/china-has-nearl.html.

5. Wikipedia contributors, "United States public debt," *Wikipedia, The Free Encyclopedia*. http://en.wikipedia.org/wiki/United_States_public_debt.

6. The Pickens Plan Community, "Did you know?" *Pickens Plan*. http://www.pickensplan.com/didyouknow/.

7. Public Utilities Commission, "California Renewable Portfolio Standard," *California Public Utilities Commission*, http://www.cpuc.ca.gov/PUC/energy/Renewables/index.htm.
8. North Carolina Solar Center, "Renewable Energy and Energy Efficiency," *Database of State Incentives for Renewables and Efficiency*, http://www.dsireusa.org/.
9. Ibid.
10. Wikipedia contributors, "Adam Smith," *Wikipedia, The Free Encyclopedia*. http://en.wikipedia.org/wiki/Adam_Smith.
11. John Wesley, *The Works of the Rev. John Wesley* (New York: J & J Harper, 1826), 7:62.
12. Infoplease, "National Voter Turnout in Federal Elections: 1960–2008," *2000–2007 Pearson Education, publishing as Infoplease.* http://www.infoplease.com/ipa/A0781453.html.
13. CBS News. "CBS Poll: 81% Say U.S. On Wrong Track: CBS/New York Times Poll Shows Americans Deeply Concerned About Economy," *CBS News.* http://www.cbsnews.com/stories/2008/04/03/opinion/polls/main3992628_page2.shtml.
14. The Great Warming Staff, "Interview with Reverend Richard Cizik," *The Great Warming.* http://www.thegreatwarming.com/revrichardcizik.html.
15. Andrew Bacevich, *The Limits of Power: The End of American Exceptionalism,* (New York: Metropolitan Books, Henry Holt and Company, LLC, 2008), 9.
16. Pew Research Center Publications, "Men or Women: Who's the Better Leader?" *Pew Research Center*, August 15, 2008, http://pewresearch.org/pubs/932/men-or-women-whos-the-better-leader.

Chapter 10

1. Mindfully.org, "Consumption by the United States," *Mindfully org*, http://www.mindfully.org/Sustainability/Americans-Consume-24percent.htm.
2. Jad Mouwad and Matthew L. Wald, "The Oil Uproar That Isn't," *The New York Times,* July 12, 2005, http://www.nytimes.com/2005/07/12/business/worldbusiness/12oil.ready.html.

3. The Late Show with David Letterman. First broadcast 8 September 2008 by CBS. Hosted by David Letterman and produced by Rob Burnett.

4. Association for the Study of Peak Oil and Gas—USA Conference, 2008, Matthew R. Simmons, "Grappling with Energy *Risk*," September 22, 2008.

5. Beth Moore, *Living beyond Yourself: Exploring the Fruit of the Spirit*, 58

6. Energy Information Administration, "Annual Energy Review 2007" *U.S. Primary Energy Consumption by Source and Sector 2007. Energy Information Administration:* http://www.eia.doe.gov/emeu/aer/pecss_diagram.html.

7. National Renewable Energy Laboratory, "Potential for Hydrogen Production from Key Renewable Resources in the United States," Department of Energy, http://www.nrel.gov/docs/fy07osti/41134.pdf, 20.

8. Energy Information Administration, "Annual Energy Review 2007" *U.S. Primary Energy Consumption by Source and Sector 2007. Energy Information Administration:* http://www.eia.doe.gov/emeu/aer/pecss_diagram.html.

9. United States Government Accountability Office, "Biofuels," Department of Energy, *GAO Highlights*, June 2007, http://www.gao.gov/highlights/d07713high.pdf.

10. Energy Information Administration, "Frequently Asked Questions-Gasoline," *Official Energy Statistics. Energy Information Administration:* http://tonto.eia.doe.gov/ask/gasoline_faqs.asp#gas_consume_year.

11. Edmunds.com, "Fuel Economy: Gas Saving Maintenance Tips," *Edmunds,* http://www.edmunds.com/advice/fueleconomy/articles/105528/article.html.

12. Lisa Fletcher and Dan Morris, "T. Boone Pickens Get Serious About the Environment," ABC News *Nightline,* July 2008, http://abcnews.go.com/print?id=5350497.

13. The Pickens Plan Community, "The Plan," *Pickens Plan.* http://www.pickensplan.com/theplan/.

14. TheOilDrum:Europe,"NaturalGas:howbigistheproblem?"December 2006, http://www.theoildrum.com/story/2006/11/27/61031/618.

15. Energy Information Administration, "World Estimated Recoverable Coal, December 31, 2005," *Energy Information Administration,* http://www.eia.doe.gov/pub/international/iea2006/table82.xls.
16. National Hydrogen Association, "H2 and You," *10 Facts,* http://www.h2andyou.org/pdf/nightLights.pdf.
17. Carbon Dioxide Information Analysis Center, "Recent Greenhouse Gas Concentrations," *Carbon Dioxide Information Analysis Center*, http://cdiac.esd.ornl.gov/pns/current_ghg.html.
18. Department of Energy Hydrogen Program, "DOE Hydrogen Program Record," *Department of Energy*, http://www.hydrogen.energy.gov/pdfs/5038_h2_cost_competitive.pdf.
19. Lara Jakes Jordan, "Gustav Revives Question: Is New Orleans Worth It?" *International Business Times,* September 2, 2008, http://www.ibtimes.com/articles/20080902/gustav-revives-question-is-new-orleans-worth-it.htm.
20. Spencer Hsu and Peter Whoriskey, "Levee Repair Costs Triple," *The Washington Post,* March 31, 2006, http://www.washingtonpost.com/wp-dyn/content/article/2006/03/30/AR2006033001912_pf.html
21. George Laughead, Jr., "City of Greensburg, Kiowa County, Kansas," *Kansas Community Network*, http://www.kansastowns.us/greensburg.html.
22. Alexander Green, "The Psychology of Optimal Experience," *Spiritual Wealth: The Road Map to a Rich Life*, March 2009, http://www.spiritualwealth.com/Archives/2008/20080808.html.

GLOSSARY

alternating current (AC): An electric current that reverses its direction at regularly recurring intervals.

albedo: The fraction of solar radiation reflected by a surface or object, often expressed as a percentage. Snow-covered surfaces have a high albedo; the albedo of soils ranges from high to low; vegetation-covered surfaces and oceans have a low albedo. The earth's albedo varies mainly through varying cloudiness, snow, ice, leaf area, and land cover changes.

alternative energy (fuel): Energy derived from nontraditional sources. An alternative to gasoline or diesel fuel that is not produced in a conventional way from crude oil, for example compressed natural gas (CNG), liquefied petroleum gas (LPG), liquefied natural gas (LNG), ethanol, methanol, and hydrogen.

AMO (Atlantic Multidecadal Oscillation): The AMO is an ongoing series of long-duration changes in the sea surface temperature of the North Atlantic Ocean, with cool and warm phases that may last for 20–40 years at a time and a difference of about 1°F between extremes. These natural changes have occurred for at least the last 1,000 years.

anode: The electrode at which oxidation (a loss of electrons) takes place. For fuel cells and other galvanic cells, the anode is the negative terminal; for electrolytic cells (where electrolysis occurs), the anode is the positive terminal.

Arctic National Wildlife Refuge (ANWR): A national wildlife refuge in northeastern Alaska. It consists of 19,049,236 acres in the Alaska North Slope region.

atmosphere: The gaseous envelope surrounding the Earth. The dry atmosphere consists almost entirely of nitrogen (78.1% volume mixing ratio) and oxygen (20.9% volume mixing ratio), together with a number of trace gases, such as argon (0.93% volume mixing ratio), helium, radiatively active greenhouse gases such as carbon dioxide (0.035% volume mixing ratio), and ozone. In addition, the atmosphere contains water vapor in highly variable amounts (but with a typical 1% volume mixing ratio). The atmosphere also contains clouds and aerosols.

auto thermal reforming (ATR): Brings together a hydrocarbon fuel, catalyst, steam, and oxygen. One feature of this system is its ability to reform many different types of fuels.

battery: An energy storage device which produces electricity by means of an electrochemical reaction. It consists of one or more electrochemical cells, each of which has all the chemicals and parts needed to produce an electric current.

benzene (C_6H_6): An aromatic hydrocarbon present in small proportion in some crude oils. Used as a solvent in the manufacture of detergents, synthetic fibers, petrochemicals, and as a component of high-octane gasoline. (This substance is carcinogenic to humans.)

biofuels: Liquid fuels and blending components produced from biomass (plant) feedstocks, used primarily for transportation.

biomass: Materials which have a biological origin, including organic material (both living and dead) from above and below ground; for example, trees, crops, grasses, tree litter, roots, animals, and animal waste.

corporate average fuel economy (CAFE): The sales-weighted average fuel economy, expressed in miles per gallon (mpg), of a manufacturer's fleet of passenger cars or light trucks with a gross vehicle weight rating (GVWR) of 8,500 lbs. or less, manufactured for sale in the United States, for any given model year. Fuel economy is defined as the average mileage traveled by an automobile per gallon of gasoline (or equivalent amount of other fuel) consumed as measured in accordance with the testing and evaluation protocol set forth by the Environmental Protection Agency (EPA).

capacity factor: The ratio of the electrical energy produced by a generating unit for the period of time considered to the electrical energy that could have been produced at continuous full power operation during the same period.

carbon cycle: All parts (reservoirs) and fluxes of carbon. The cycle is usually thought of as four main reservoirs of carbon interconnected by pathways of exchange. The reservoirs are the atmosphere, terrestrial biosphere (usually includes freshwater systems), oceans, and sediments (includes fossil fuels). The annual movements of carbon, the carbon exchanges between reservoirs, occur because of various chemical, physical, geological, and biological processes. The ocean contains the largest pool of carbon near the surface of the Earth, but most of that pool is not involved with rapid exchange with the atmosphere.

carbon dioxide (CO_2): A naturally-occurring, colorless, odorless, non-poisonous gas. A by-product of burning fossil fuels and biomass, as well as land-use changes and other industrial processes, it's the principal anthropogenic greenhouse gas affecting the Earth's radiative balance and the reference gas against which other greenhouse gases are measured. It has a Global Warming Potential (GWP) of 1.

carbon sequestration: The uptake and storage of carbon. Trees and plants, for example, absorb carbon dioxide, release the oxygen, and store the carbon. Fossil fuels were, at one time, biomass and continue to store the carbon until burned.

carbon sink: A reservoir that absorbs or takes up released carbon from another part of the carbon cycle. The four sinks, which are regions of the Earth within which carbon behaves in a systematic manner, are the atmosphere, terrestrial biosphere (usually including freshwater systems), oceans, and sediments (including fossil fuels).

catalyst: A chemical substance that increases the rate of a reaction without being consumed; after the reaction it can potentially be recovered from the reaction mixture chemically unchanged. The catalyst lowers the activation energy required, allowing the reaction to proceed more quickly or at a lower temperature. In a fuel cell, the catalyst facilitates the reaction of oxygen and hydrogen. It is usually made of platinum powder very thinly coated onto carbon paper or cloth. The catalyst is rough and porous so that the maximum surface area of the platinum can be exposed to the hydrogen or oxygen. The platinum-coated side of the catalyst faces the membrane in the fuel cell.

catalytic cracking: The refining process of breaking down the larger, heavier, and more complex hydrocarbon molecules into simpler and lighter molecules. Catalytic cracking is accomplished by the use of a catalytic agent and is an effective process for increasing the yield of gasoline from crude oil. Catalytic cracking processes fresh feeds and recycled feeds.

catalytic hydro-cracking: A refining process that uses hydrogen and catalysts with relatively low temperatures and high pressures for converting middle boiling or residual material to high octane gasoline, reformer charge stock, jet fuel, and /or high grade fuel oil. The process uses one or more catalysts, depending on product output, and can handle high sulfur feedstocks without prior desulfurization.

catalytic hydro-treating: A refining process for treating petroleum fractions from atmospheric or vacuum distillation units (e.g., naphthas, middle distillates, reformer feeds, residual fuel oil, and heavy gas oil) and other petroleum (e.g., cat cracked naphtha, coker naphtha, gas oil, etc.) in the presence of catalysts and substantial quantities of hydrogen. Hydro-treating includes desulfurization, removal of substances (e.g., nitrogen compounds) that deactivate catalysts, conversion of olefins to paraffins to reduce gum formation in gasoline, and other processes to upgrade the quality of the fractions.

cathode: The electrode at which reduction (a gain of electrons) occurs. For fuel cells and other galvanic cells, the cathode is the positive terminal; for electrolytic cells (where electrolysis occurs), the cathode is the negative terminal.

combined heat and power (CHP): A plant designed to produce both heat and electricity from a single heat source. Note: This term replaces the term "cogenerator," used by EIA in the past. CHP better describes the facilities because some plants do not produce heat and power in a sequential fashion and, as a result, do not meet the legal definition of cogeneration specified in the Public Utility Regulatory Policies Act (PURPA).

climate change: Climate change refers to any significant change in measures of climate (such as temperature, precipitation, or wind) lasting for an extended period (decades or longer). Climate change may result from: natural factors, such as changes in the sun's intensity or slow changes in the Earth's orbit around the sun; natural processes within the climate system (e.g. changes in ocean circulation); human activities which change the atmosphere's composition (e.g. through burning fossil fuels) and the land surface (e.g. deforestation, reforestation, urbanization, desertification, etc.)

CIGS (copper indium gallium diselenide): A polycrystalline thin-film photovoltaic material incorporating gallium (CIGS) and/or sulfur.

coal gasification: The process of converting coal into gas. The basic process involves crushing coal to a powder, then heating it in the presence of steam and oxygen to produce a gas. The gas is then refined to reduce sulfur and other impurities. The gas can be used as a fuel or processed further and concentrated into chemical or liquid fuel.

coal to liquids (CTL): A chemical process that converts coal into clean-burning liquid hydrocarbons, such as synthetic crude oil and methanol.

conservation: Managing natural resources to prevent loss or waste.

conventional oil and natural gas production: Crude oil and natural gas that is produced by a well drilled into a geologic formation in which the reservoir and fluid characteristics permit the oil and natural gas to readily flow to the bore hole of the drilled well.

direct current (DC): Electric current in which electrons flow in one direction only. Opposite of alternating current.

deforestation: Practices or processes which result in the conversion of forested lands for non-forest uses. It's often cited as a major cause of the enhanced greenhouse effect for two reasons: 1) the burning or decomposition of the wood releases carbon dioxide; and 2) absence of trees which once removed carbon dioxide from the atmosphere in the process of photosynthesis.

distributed generation: Any small-scale power generation technology which provides electric power at or closer to the customer's site than centrally-sited generation stations; a generator located close to the particular load it's intended to serve.

efficiency: A value-based, philosophical concept. Two different concepts of energy efficiency are discussed, a technical and a more broad, subjective concept. In the technical concept, increases in energy efficiency take place when either energy inputs are reduced for a given level of service or there are increased or enhanced services for a given amount of energy inputs. In the more subjective concept, energy efficiency is the relative thrift or extravagance with which energy inputs are used to provide goods or services.

electric grid: A system of synchronized power providers and consumers connected by transmission and distribution lines, and operated by one or more control centers. In the continental United States, the electric power grid consists of three systems: the Eastern Interconnect, the Western Interconnect, and the Texas Interconnect.

electrolysis: A process that passes electricity through an electrolytic solution or other appropriate medium to cause a reaction that breaks chemical bonds, e.g., electrolysis of water to produce hydrogen and oxygen.

electrolyzer: A device that uses a flow of electrons to break a compound into its constituent elements.

energy: The capacity for doing work as measured by the capability of doing work (potential energy) or the conversion of this capability to motion (kinetic energy). Energy has several forms, some of which can be easily converted to another form useful for work. Most of the world's convertible energy comes from fossil fuels burned to produce heat, which is then used as a transfer medium to mechanical or other means in order to accomplish tasks. Electrical energy is usually measured in kilowatt hours (kWh), while heat energy is usually measured in British thermal units (Btu).

energy carrier: Something that can store and deliver energy in a usable form (note: energy carriers such as electricity and hydrogen are not energy sources—they must be produced using energy sources).

energy content: Amount of energy for a given weight of fuel.

energy density: Amount of potential energy in a given measurement of fuel.

energy intensity: The ratio of energy consumption to a measure of the demand for services (e.g., number of buildings, total floorspace, floorspace hours, number of employees, or constant dollar value of gross domestic product for services).

ethanol (CH_3CH_2OH): An alcohol containing two carbon atoms, it's a clear, colorless liquid and the same alcohol found in beer, wine, and whiskey. Ethanol can be produced from cellulosic materials, or by fermenting a sugar solution with yeast.

externalities: Benefits or costs, generated as a byproduct of an economic activity, that do not accrue to the parties involved in the activity. Environmental externalities are benefits or costs that manifest themselves through changes in the physical or biological environment.

fission: The process whereby an atomic nucleus of appropriate type, after capturing a neutron, splits into (generally) two nuclei of lighter elements, with the release of substantial amounts of energy and two or more neutrons.

flared natural gas: Gas disposed of by burning in flares, usually at the production sites or at gas processing plants.

fossil fuel: An energy source formed in the Earth's crust from decayed organic material. The common fossil fuels are petroleum, coal, and natural gas.

fuel cell: A device capable of generating an electrical current by converting the chemical energy of a fuel (e.g., hydrogen) directly into electrical energy. Fuel cells differ from conventional electrical cells in that the active materials, such as fuel and oxygen, are not contained within the cell but are supplied from outside. It does not contain an intermediate heat cycle, as do most other electrical generation techniques.

fuel cell poisoning: The lowering of a fuel cell's efficiency due to impurities in the fuel binding to the catalyst.

fusion: When the nuclei of atoms are combined (fused) together. The sun combines the nuclei of hydrogen atoms into helium atoms in a process called fusion. Energy from the nuclei of atoms, called *nuclear energy* is released from fusion.

gallon gasoline equivalent (gge): the energy content equivalent to one gallon of gasoline (114,320 Btu).

gas-cooled fast reactor (GFR): The Gas-Cooled Fast Reactor (GFR) system features a fast-spectrum, helium-cooled reactor (or super-critical CO_2) and closed fuel cycle. The main characteristics of the GFR are: a self-generating core with a fast neutron spectrum, robust refractory fuel, high operating temperature, direct energy conversion with a gas turbine, and full actinide recycling.

Gen IV: An entire next-generation nuclear energy production system, including the nuclear fuel cycle front and back end, the reactor, the power conversion equipment and its connection to the distribution system for electricity, hydrogen, process heat or fresh water, and the infrastructure for manufacture and deployment of the plant.

geothermal energy: Hot water or steam extracted from geothermal reservoirs in the earth's crust. Water or steam extracted from geothermal reservoirs can be used for geothermal heat pumps, water heating, or electricity generation.

greenhouse effect: Trapping and build-up of heat in the atmosphere (troposphere) near the Earth's surface. Some of the heat flowing back toward space from the Earth's surface is absorbed by water vapor, carbon dioxide, ozone, and several other gases in the atmosphere and then radiated back toward the Earth's surface. If atmospheric concentrations of these greenhouse gases rise, the average temperature of the lower atmosphere will gradually increase.

greenhouse gas (GHG): Any gas that absorbs infrared radiation in the atmosphere. Greenhouse gases include, but are not limited to, water vapor, carbon dioxide (CO_2), methane (CH4), nitrous oxide (N_2O), chlorofluorocarbons (CFCs), hydrochlorofluorocarbons (HCFCs), ozone (O_3), hydrofluorocarbons (HFCs), perfluorocarbons (PFCs), and sulfur hexafluoride (SF_6).

high temperature gas reactor (HTGR) or very-high-temperature reactor (VHTR): A graphite-moderated, helium-cooled reactor with a once-through uranium fuel cycle. It supplies heat with high core outlet temperatures which enables applications such as hydrogen production or process heat for the petrochemical industry or others.

hydrides: Chemical compounds formed when hydrogen gas reacts with metals. Used for storing hydrogen gas.

hydrocarbon (HC): An organic compound containing carbon and hydrogen, usually derived from fossil fuels, such as petroleum, natural gas, and coal.

hydrogen (H_2): Hydrogen (H) is the most abundant element in the universe but it is generally bonded to another element. Hydrogen gas (H_2) is a diatomic gas composed of hydrogen atoms and is colorless and odorless. Hydrogen is flammable when mixed with oxygen over a wide range of concentrations.

integrated gasification combined cycle (IGCC): A combination of two leading technologies. The first technology is called coal gasification, which uses coal to create a clean-burning gas (syngas). The second technology is called combined-cycle, which is the most efficient method of producing electricity commercially available today.

internal combustion engine (ICE): An engine that converts the energy contained in a fuel inside the engine into motion by combusting the fuel. Combustion engines use the pressure created by the expansion of combustion product gases to do mechanical work.

Intergovernmental Panel on Climate Change (IPCC): Established jointly by the United Nations Environment Programme and the World Meteorological Organization in 1988, the IPCC assesses information in the scientific and technical literature related to all significant components of the issue of climate change. The IPCC draws upon hundreds of the world's expert scientists as authors, and thousands as expert reviewers. Leading experts on climate change and environmental, social, and economic sciences from some 60 nations have helped the IPCC to prepare periodic assessments of the scientific underpinnings for understanding global climate change and its consequences. With its capacity for reporting on climate change, its consequences, and the viability of adaptation and mitigation measures, the IPCC also functions as the official advisory body to the world's governments on the state of the science of the climate change issue.

intermittency: The variable nature of the timing of the generation of electrical power from a wind turbine, reflecting that the amount of wind varies by a number of factors, including season, time of day, and the regional topography. Also, power generating systems whose output is generally inconsistent or variable, such as those that rely on solar or wind sources.

liquefied natural gas (LNG): Natural gas (primarily methane) that has been liquefied by reducing its temperature to -260 degrees Fahrenheit at atmospheric pressure.

molten carbonate fuel cell (MCFC): A type of fuel cell that contains a molten carbonate electrolyte. Carbonate ions (CO_3^{-2}) are transported from the cathode to the anode. Operating temperatures are typically near 650°C.

methane: A colorless, flammable, odorless hydrocarbon gas (CH_4) which is the major component of natural gas. It is also an important source of hydrogen in various industrial processes. Methane is a greenhouse gas.

Mount Pinatubo: A volcano in the Philippine Islands, the eruption of Mount Pinatubo in 1991 ejected enough particulate and sulfate aerosol matter into the atmosphere to block some of the incoming solar radiation from reaching Earth's atmosphere. This effectively cooled the planet from 1992 to 1994, masking the warming that had been occurring for most of the 1980s and 1990s.

MTBE (Methyl Tertiary Butyl Ether) ($(CH_3)_3COCH_3$: An ether intended for gasoline blending. See also *oxygenates*.

natural gas hydrates: Solid, crystalline, wax-like substances composed of water, methane, and usually a small amount of other gases, with the gases being trapped in the interstices of a water-ice lattice. They form beneath permafrost and on the ocean floor under conditions of moderately high pressure and at temperatures near the freezing point of water.

natural gas liquids (NGL): Those hydrocarbons in natural gas that are separated from the gas as liquids through the process of absorption, condensation, adsorption, or other methods in gas processing or cycling plants. Generally such liquids consist of propane and heavier hydrocarbons and are commonly referred to as lease condensate, natural gasoline, and liquefied petroleum gases. Natural gas liquids include natural gas plant liquids (primarily ethane, propane, butane, and isobutane).

National Renewable Energy Laboratory (NREL): Located in Golden, Colorado. Leading center for energy innovation for the U.S. Department of Energy.

net metering: An electricity policy typically for owners of small, renewable energy facilities. "Net" in this context means the deduction of any energy outflows from metered energy inflows.

nuclear energy (nuclear power): Electricity generated by the use of the thermal energy released from the fission of nuclear fuel in a reactor.

oil sands: Also called "tar sands," a sedimentary material composed primarily of sand, clay, water, and organic constituents known as bitumen.

oil shale: A sedimentary rock containing kerogen, a solid organic material.

Organization of the Petroleum Exporting Countries (OPEC): An intergovernmental organization which coordinates and unifies petroleum policies among member countries. Created at the Baghdad Conference on September 10–14, 1960, by Iran, Iraq, Kuwait, Saudi Arabia and Venezuela, the five founding members were later joined by nine other members: Qatar (1961); Indonesia (1962); Libya (1962); United Arab Emirates (1967); Algeria (1969); Nigeria (1971); Ecuador (1973–1992, 2007); Gabon (1975–1994) and Angola (2007).

oxygenates: Substances which, when added to gasoline, increase the amount of oxygen in that gasoline blend. Ethanol, Methyl Tertiary Butyl Ether (MTBE), Ethyl Tertiary Butyl Ether (ETBE), and methanol are common oxygenates.

phosphoric acid fuel cell (PAFC): A type of fuel cell in which the electrolyte consists of concentrated phosphoric acid (H_3PO_4). Protons (H^+) are transported from the anode to the cathode. The operating temperature range is generally 160–220°C.

polymer electrolyte membrane (PEM): A fuel cell incorporating a solid polymer membrane used as its electrolyte. Protons (H^+) are transported from the anode to the cathode. The operating temperature range is generally 60–100°C.

power: The rate of producing, transferring, or using energy, most commonly associated with electricity. Power is measured in watts and often expressed in kilowatts (kW) or megawatts (MW). Also known as "real" or "active" power.

power quality: The difference between the quality of electricity at an electrical outlet and the quality of the electricity required to reliably operate an appliance. Faulty operation or damage to the appliance could result if power quality requirements are not met.

rate base: The value of property upon which a utility is permitted to earn a specified rate of return as established by a regulatory authority. The rate base generally represents the value of property used by the utility in providing service and may be calculated by any one or a combination of the following accounting methods: fair value, prudent investment, reproduction cost, or original cost. Depending on which method is used, the rate base includes cash, working capital, materials and supplies, deductions for accumulated provisions for depreciation, contributions in aid of construction, customer advances for construction, accumulated deferred income taxes, and accumulated deferred investment tax credits.

regenerative fuel cell: A fuel cell that produces electricity from hydrogen and oxygen and can use electricity from solar power or some other source to divide the excess water into oxygen and hydrogen fuel to be re-used by the fuel cell.

Renewable energy credits or certificates (REC): The tradable commodity certificates representing the environmental aspect of electricity generated by renewable energy unbundled from the commodity electricity.

reformer: Device that extracts pure hydrogen from hydrocarbon fuels.

renewable energy: Energy resources which are naturally replenishing but flow-limited. They are virtually inexhaustible in duration but limited in the amount of energy that is available per unit of time. Renewable energy resources include: biomass, hydro, geothermal, solar, wind, ocean thermal, wave action, and tidal action.

reliability: Electric system reliability has two components—adequacy and security. Adequacy is the ability of the electric system to supply to aggregate electrical demand and energy requirements of the customers at all times, taking into account scheduled and unscheduled outages of system facilities. Security is the ability of the electric system to withstand sudden disturbances, such as electric short circuits or unanticipated loss of system facilities. The degree of reliability may be measured by the frequency, duration, and magnitude of adverse effects on consumer services.

renewable portfolio standards (RPS): A policy set by federal or state governments requiring that a fixed percentage of electricity used by utilities be derived from designated renewable sources.

residence time: The average time spent in a reservoir by an individual atom or molecule. With respect to greenhouse gases, residence time usually refers to how long a particular molecule remains in the atmosphere.

steam methane reforming (SMR): The process for reacting a hydrocarbon fuel, such as natural gas, with steam to produce hydrogen as a product. This is a common method for bulk hydrogen generation.

solid oxide fuel cell (SOFC): A type of fuel cell in which the electrolyte is a solid, nonporous metal oxide, typically zirconium oxide (ZrO_2) treated with Y_2O_3, and O^{-2} is transported from the cathode to the anode. Any CO in the reformate gas is oxidized to CO_2 at the anode. Temperatures of operation are typically 800–1,000°C.

spinning reserve: That reserve generating capacity running at a zero load and synchronized to the electric system.

stoichiometric: The ideal combustion process where fuel is burned completely. A complete combustion is a process burning all the carbon (C) to (CO_2), all the hydrogen (H) to (H_2O) and all the sulphur (S) to (SO_2).

stratosphere: The region of the upper atmosphere extending from the tropopause (8 to 15 kilometers altitude) to about 50 kilometers. Its thermal structure, determined by its radiation balance, is generally very stable with low humidity.

sulfur: A yellowish nonmetallic element, sometimes known as brimstone. It is present at various levels of concentration in many fossil fuels whose combustion releases sulfur compounds considered harmful to the environment. Some of the most commonly used fossil fuels are categorized according to their sulfur content, with lower sulfur fuels usually selling at a higher price. *Note:* No. 2 Distillate fuel is currently reported as having either a 0.05 percent or lower sulfur level for on-highway vehicle use or a greater than 0.05 percent sulfur level for off-highway use, home heating oil, and commercial and industrial uses. Residual fuel, regardless of use, is classified as having either no more than 1 percent sulfur or greater than 1 percent sulfur. Coal is also classified as being low- sulfur at concentrations of 1 percent or less or high-sulfur at concentrations greater than 1 percent.

volt (V): The volt is the International System of Units (SI) measure of electric potential or electromotive force. A potential of one volt appears across a resistance of one ohm when a current of one ampere flows through that resistance.

volatile organic compounds (VOCs): Organic compounds that participate in atmospheric photochemical reactions.

watt (W): The unit of electrical power equal to one ampere under a pressure of one volt. A watt is equal to 1/746 horsepower.

water vapor: The most abundant greenhouse gas, it's the water present in the atmosphere in gaseous form. Water vapor is an important part of the natural greenhouse effect. While humans do not significantly increase its concentration, it contributes to the enhanced greenhouse effect because the warming influence of greenhouse gases leads to a positive water vapor feedback. In addition to its role as a natural greenhouse gas, water vapor plays an important role in regulating the temperature of the planet because clouds form when excess water vapor in the atmosphere condenses to form ice and water droplets and precipitation.

Watt-hour (Wh): The electrical energy unit of measure equal to one watt of power supplied to, or taken from, an electric circuit steadily for one hour.

weather: Atmospheric condition at any given time or place. It is measured in terms of such things as wind, temperature, humidity, atmospheric pressure, cloudiness, and precipitation. In most places, weather can change from hour-to-hour, day-to-day, and season-to-season. Climate in a narrow sense is usually defined as the average weather, or more rigorously, as the statistical description in terms of the mean and variability of relevant quantities over a period of time ranging from months to thousands or millions of years. The classical period is 30 years, as defined by the World Meteorological Organization (WMO). These quantities most often include surface variables such as temperature, precipitation, and wind. Climate in a wider sense is the state, including a statistical description, of the climate system. A simple way of remembering the difference is that climate is what you expect (e.g. cold winters) and weather is what you get (e.g. a blizzard).

zero emission vehicle (ZEV): A vehicle that produces no air emissions from its fueling or operation. California regulations require that in 2003, 10% of the vehicles sold in California by major automakers be ZEV or ZEV-equivalent. California has established a comprehensive program for determining this equivalency.

To order additional copies of this title call:
1-877-421-READ (7323)
or please visit our Web site at
www.winepressbooks.com

If you enjoyed this quality custom-published book,
drop by our Web site for more books and information.

www.winepressgroup.com
"Your partner in custom publishing."